IMPLEMENTING TOTAL QUALITY

DAVID L. GOETSCH

STANLEY DAVIS

Prentice Hall

Englewood Cliffs, New Jersey   Columbus, Ohio

Library of Congress Cataloging-in-Publication Data
Goetsch, David L.
    Implementing total quality/David L. Goetsch, Stanley Davis.
       p.   cm.
    Includes bibliographical references and index.
    ISBN 0-02-344224-7
    1. Total quality management. I. Davis, Stanley
  II. Title
  HD62. 15.G638 1995   94-34208
  658.5'62—dc20   CIP

Editor: Stephen Helba
Production Editor: Christine M. Harrington
Cover Designer: Brian Deep
Production Buyer: Laura Messerly
Electronic Text Management: Marilyn Wilson Phelps, Matthew Williams, Jane Lopez, Karen L. Bretz

This book was set in Clearface by Prentice Hall and was printed and bound
by Courier Corp. The cover was printed by Phoenix Color Corp.

© 1995 by Prentice-Hall, Inc.
A Simon & Schuster Company
Englewood Cliffs, New Jersey 07632

Printed in the United States of America

10 9 8 7 6 5 4 3 2 1

ISBN: 0-02-344224-7

Prentice-Hall International (UK) Limited, *London*
Prentice-Hall of Australia Pty. Limited, *Sydney*
Prentice-Hall of Canada, Inc., *Toronto*
Prentice-Hall Hispanoamericana, S. A., *Mexico*
Prentice-Hall of India Private Limited, *New Delhi*
Prentice-Hall of Japan, Inc., *Tokyo*
Simon & Schuster Asia Pte. Ltd., *Singapore*
Editora Prentice-Hall do Brasil, Ltda., *Rio de Janeiro*

PREFACE

WHY WAS THIS BOOK WRITTEN AND FOR WHOM?

This book was written as a supplement to the authors' earlier book, *Introduction to Total Quality: Quality, Productivity, and Competitiveness.* That book was written in response to the need for a practical teaching resource that encompasses all of the various elements of the Total Quality approach and pulls them together in a coherent format that allows the reader to understand both the big picture and the specific details of Total Quality. Chapter 18 of the book explained a 20-step process the authors developed for implementing Total Quality in any organization.

Response to *Introduction to Total Quality* overall, and to Chapter 18 in particular, was so enthusiastic that the need for a much more in-depth treatment of how to implement Total Quality quickly became apparent. *Implementing Total Quality* was written in response to that need. It is intended for use as a step-by-step guide to implementing Total Quality in any organization. Consequently, users should have either completed the earlier book, *Introduction to Total Quality: Quality, Productivity, and Competitiveness,* or should be knowledgeable about Total Quality as the result of other experiences.

ORGANIZATION OF THIS BOOK

This book begins with a comprehensive Introduction that provides an overview of Total Quality. Successive chapters explain each step in the 20-step implementation process. Each chapter begins with a list of major topics and ends with a chapter summary, key terms list, review questions, endnotes, and two case studies. One case study illustrates the right way to carry out the implementation step explained in the chapter; the other illustrates the wrong way. The case studies build on one another from chapter to chapter so that readers can follow two companies through the entire implementation process and learn how one succeeds and one fails.

ABOUT THE AUTHORS

David L. Goetsch is Provost of the joint campus of the University of West Florida and Okaloosa-Walton Community College in Fort Walton Beach, Florida. He also adminsters the state of Florida's Center for Manufacturing Competitiveness that is located on this campus. In addition, Dr. Goetsch is President of the Institute for Corporate Competitiveness, a private consulting firm dedicated to the continual improvement of organizational competitiveness. Dr. Goetsch is co-founder of The Quality Institute, a partnership of the University of West Florida, Okaloosa-Walton Community College, and the Okaloosa Economic Development Council. He currently serves on the executive board of the Institute.

Stanley Davis was a manufacturing executive with Harris Corporation until his retirement in 1991. He is currently the managing director of The Quality Institute and a well-known expert in the areas of implementing total quality, statistical process control, just-in-time manufacturing, and benchmarking.

CONTENTS

Overview of Total Quality

The Total Quality concept as a way to continually improve competitiveness began to gain wide acceptance in the United States in the late 1980s and early 1990s. However, individual elements of the concept, such as the use of statistical data, teamwork, and employee involvement, have been used by visionary organizations for years. The pulling together and coordinated use of these and other previously disparate elements gave birth to the comprehensive concept known as Total Quality. This introduction provides an overview of the concept that lays the foundation for the study of how to implement Total Quality in any organization.

WHAT IS QUALITY?

In order to understand Total Quality, one must first understand quality. Although few consumers can define quality as a concept, they know it when they see it. This makes the critical point that quality is in the eye of the beholder, and the beholder is the customer. This is why customer-defined quality is essential to competitiveness.

People deal with the issue of quality every day. Consumers are concerned with quality whenever they shop for groceries, eat in a restaurant, and make purchases. Perceived quality is a major factor by which people make distinctions in the marketplace. Whether consumers openly articulate these distinctions or simply have them in mind, they all apply a number of criteria when making a purchase. The extent to which the purchase meets these criteria determines its quality in the consumer's eyes.

One way to understand quality as a consumer-driven concept is to consider the example of eating out at a restaurant. How is the quality of the restaurant judged? Most people apply such criteria as service, response time, food preparation, atmosphere, price, and selection.

This example speaks to one aspect of quality: results. Does the product or service meet or exceed customer expectations? While this is a critical aspect of quality, it is not the only one. Total Quality is a much broader concept that encompasses not only the results, but also the quality of the people involved and the quality of the processes employed. According to Stephen Uselac,

> There is little agreement on what constitutes quality. In its broadest sense, quality is an attribute of a product or service that can be improved. Most people associate quality with a product or service. Quality IS NOT only products and services but also includes PROCESSES, ENVIRONMENT, and PEOPLE.[1]

Quality has been defined in a variety of ways by a number of different people and organizations. Consider the following examples:

- Fred Smith, CEO of Federal Express, defines quality as "performance to the standard expected by the customer. . . . ."[2]
- The General Services Administration defines quality as "meeting the customer's needs the first time and every time."[3]
- Boeing defines quality as "providing our customers with products and services that consistently meet their needs and expectations."[4]
- The U.S. Department of Defense defines quality as "doing the right thing right the first time, always striving for improvement, and always satisfying the customer."[5]

In his landmark book, *Out of the Crisis,* W. Edwards Deming has this to say about quality:

> Quality can be defined only in terms of the agent. Who is the judge of quality? In the mind of the production worker, he produces quality if he can take pride in his work. Poor quality, to him, means loss of business, and perhaps of his job. Good quality, he thinks, will keep the company in business. Quality to the plant manager means to get the numbers out and to meet specifications. His job is also, whether he knows it or not, continual improvement of leadership.[6]

Deming goes on to make the point that quality has many different criteria and that these criteria change continually.[7] To complicate matters further, different people value the different criteria differently. For this reason, it is important to measure consumer

TOTAL QUALITY TIP

Quality Encompasses both Substance and Image

"Quality encompasses every aspect of your firm and is actually an emotional experience for the customer. Customers want to feel good about their purchases, to feel that they have gotten the best value. They want to know their money has been well spent, and they take pride in their association with a company with a high quality image."[8]

Perry L. Johnson

preferences and to remeasure them frequently. Deming gives an example of the criteria that are important to him in selecting paper:[9]

- It is not slick and therefore takes pencil or ink well.
- Writing on the back does not show through.
- It fits into a three-ring notebook.
- It is available at most stationery stores and therefore is easily replenished.
- It is reasonably priced.

Each of these preferences represents a variable the manufacturer can measure and use to continually improve decision making.

Deming is well known for helping Japan rise up out of the ashes of World War II to become a world-class industrial power, and for his belief that 85 percent of workplace problems are caused by management. Deming's contributions to the quality movement are explained in greater depth later in this chapter.

Although there is no universally accepted definition of quality, the various definitions have enough similarity to extract the following common elements:

- Quality involves meeting or exceeding customer expectations.
- Quality applies to products, services, people, processes, and environments.
- Quality is an ever-changing state—what is considered quality today may not be good enough to be considered quality tomorrow.

With these common elements identified, the following definition of quality is provided:

> *"Quality is a dynamic state associated with products, services, people, processes, and environments that meet or exceed current expectations."*

TOTAL QUALITY TIP

Consumer Preference Studies

"The purpose of studies in consumer preference is to adjust the product to the public, rather than, as in advertising, to adjust the public to the product."[10]

Irwin Bross

It can be instructive to examine the individual elements of this definition.

The *dynamic state* element speaks to the fact that what is considered quality can and often does change as time passes and circumstances are altered. For example, gas mileage is an important criterion in judging the quality of modern automobiles. However, in the days when gasoline cost 20 cents per gallon, consumers were more likely to be concerned with an automobile's horsepower, engine size, and acceleration rate than with gas mileage.

The *products, services, people, processes, and environments* element is critical. It makes the point that quality applies not just to the products and services provided, but also to the people and processes that provide them and the environments in which they are provided. There is an important reason for this; it's called longevity. In the short term, two competitors who focus on continual improvement might produce a product of comparable quality. But the competitor who looks beyond just the quality of the finished product and also focuses on the continual improvement of the people who produce the product, the processes they use, and the environment in which they work will win in the long run, and frequently in the short run. This is because quality products are produced most consistently by quality organizations. This phenomenon is illustrated Case Study I–1.

TOTAL QUALITY DEFINED

Just as there are different definitions for quality, there are different definitions for Total Quality. For example, the U.S. Department of Defense uses the term Total Quality Management or TQM and defines it as follows:

> TQM consists of continuous improvement activities involving everyone in the organization—managers and workers—in a totally integrated effort toward improving performance at every level. This improved performance is directed toward satisfying such cross-functional goals as quality, cost, schedule, mission need, and suitability. TQM integrates fundamental management techniques, existing improvement efforts, and technical tools under a disciplined approach focused on continual process improvement. The activities are ultimately focused on increased customer/user satisfaction."[11]

============ **CASE STUDY I–1** ==

Winning and Longevity

A professional baseball team set its sights on winning the World Series. The team owner wanted to win big and win fast. Consequently, the team sank all of its resources into trading for the best players in the league. It was able to obtain enough good players that within two seasons the team was the World Series champion. However, the team had committed such a high percentage of its financial resources to players' salaries that other important aspects of the team began to suffer. Its stadium quickly fell into such a state of disrepair that fans began to stay home. Its training facilities also declined, causing discontent among the players. Insufficient money was left over for coaches' salaries, so most of the good coaches accepted better offers from other teams. In short, by focusing so intently on its desired end result, this organization neglected the other important aspects of building a competitive team. As a result, their World Series championship was a short-lived, once-in-a-lifetime victory. Before long, the team's crumbling infrastructure sent them tumbling to the bottom of their division. Without the people, processes, and environment to turn the situation around, the team was eventually sold at a loss and moved to another city.

Although there is much of value in this definition, it would be difficult to repeat should one be asked to define Total Quality. This is the case with many of the definitions that have been set forth by various organizations and individuals. A problem inherent in most definitions is that they attempt to incorporate both what Total Quality is and how it is achieved. By separating the two, a more workable definition can be achieved. For the purpose of this book, Total Quality is defined as shown in Figure I–1.

The first part of this definition describes WHAT Total Quality is. The second part explains HOW it is achieved. In the case of Total Quality, the HOW is critical because this is what separates the Total Quality approach from other approaches to doing business.

The TOTAL in Total Quality indicates a concern for quality in the broadest sense, or what has come to be known as the Big Q. The Big Q refers to quality of products, services, people, processes, and environments. Correspondingly, the Little Q refers to a narrower concern that focuses on the quality of one of these elements or individual quality criteria within a given element.

What's in a Name?

People seem to have an inherent need to label concepts with three- or four-letter acronyms, such as MBO for Management by Objectives and MBWA for Management by Walking Around. Predictably, the Total Quality movement has fallen prey to this practice. Acronyms such as TQL for Total Quality Leadership, TQC for Total Quality Control, and TQM for Total Quality Management are used by advocates of Total Quality. Since all of

What It Is
Total quality is an approach to doing business that attempts to maximize the competitiveness of an organization through the continual improvement of the quality of its products, services, people, processes, and environments.

How It Is Achieved
The total quality approach has the following characteristics:
- Customer focus (internal and external)
- Obsession with quality
- Scientific approach to decision making and problem solving
- Long-term commitment
- Teamwork
- Continual process improvement
- Education and training
- Freedom through control
- Unity of purpose
- Employee involvement and empowerment

Figure I–1
Total Quality Defined

these acronyms describe the concept set forth in Figure I–1, the ongoing battle for acronym supremacy seems to have more to do with the marketing strategies of proponents of Total Quality than with actual differences in meaning or even emphasis.

The use of acronyms has a number of other pitfalls:

- Acronyms given to management innovations tend to last only until the next innovation gives birth to a new acronym. Total Quality is not simply a management innovation—it is a whole new way of doing business.

- Acronyms rarely do justice to the whole range of a concept. For example, the acronym TQL makes the concept appear to be a leadership issue only. Although leadership is a critical element of Total Quality, successful Total Quality requires more than just leadership. Similarly, the acronym TQM implies that the concept is just a management issue, when in fact much of what is required in order to successfully implement Total Quality is worker dependent rather than management oriented.

- Too often management concepts that are labeled with acronyms are viewed by practitioners as gimmicks that will go away as quickly as they came, and that therefore can be ignored or waited out. Total Quality is not a management gimmick. Rather, it is a completely different way of doing business that requires long-term commitment and unity of purpose.

For these reasons, in this book the authors avoid acronyms and use the term *Total Quality* to refer to the concept about which the book was written.

How Is Total Quality Different?

The elements that distinguish Total Quality from traditional ways of doing business can be found in how it is achieved. The distinctive characteristics of Total Quality are as follows:

- Customer focus (internal and external)
- Obsession with quality
- Use of the scientific approach in decision making and problem solving
- Long-term commitment
- Teamwork
- Employee involvement and empowerment
- Continual improvement
- Bottom-up education and training
- Freedom through control
- Unity of purpose

These characteristics are explained later in this chapter.

What Total Quality Is and Is Not

When global competition became a reality for even small and medium-sized organizations in the mid 1980s, some forward-looking decision-makers began to look to the adoption of Total Quality as a way to survive. Unfortunately, too many saw it as a "trick play" sent in to win the game in the last minute rather than an entirely new way to play an entirely new game. An example is given in Case Study I–2.

Total Quality is neither a magic cure nor a quick fix. Consequently, organizations that have approached Total Quality as such have failed to implement it successfully. In many cases, the implementation of Total Quality has paralleled that of the "new math," an innovative approach to teaching mathematics developed in the 1960s. Pilot studies involving teachers who understood and believed in the approach produced encouraging results among students who had previously experienced great difficulty in grasping mathematical concepts. Encouraged by these results, proponents rushed new math materials to the market, and many school districts adopted the concept as a way to boost the declining performance of students in mathematics. Unfortunately, the task of teaching the new math often fell to teachers who were accustomed to the old approach and who neither understood nor believed in the new one. Their half-hearted attempts to teach new math using old math techniques guaranteed its failure.

Too often organizations take a similar approach to implementing Total Quality. This is what happened to the Crestfield Precast Corporation in Case Study I–2. Total Quality is not a quick fix, a trick play, or a gimmick. Rather, it is a whole new approach to doing business that requires a complete revamping of traditional management paradigms, long-term commitment, unity of purpose, and specialized training. This is illustrated in Case Study I–3.

—————————————— CASE STUDY I–2 ——————————————

A Half-Hearted Attempt at Total Quality

John Benning* was CEO of Crestfield Precast Corporation,* a medium-sized manufac-turing company that produced precast concrete products for the construction industry. When Crestfield's business base dropped from $9 million to $4 million in less than five years, Benning became desperate to stop the downhill slide. Having heard and read about the total quality approach that other companies, including several of his competi-tors, were using, Benning decided to try it. He hired a high-powered, high-priced con-sultant to provide a three-day seminar on total quality for all of Crestfield's employees. At the end of the seminar, Benning told his workforce, "Now that you know how to do total quality, get started."

Predictably, with no more direction than this, things did not go too well. Within two weeks, Crestfield was in a state of confusion. Quality circles were nothing more than daily complaint sessions. There was no real evidence of management commitment. Con-sequently, supervisors were reluctant to solicit employee input and employees were even more reluctant to give it. Rather than achieving unity of purpose, the attempt brought Crestfield ambiguity and confusion.

By the end of the first month, Benning had had enough. He called his top managers together and, in a fit of frustration, yelled, "Stop this total quality nonsense right now!" Crestfield filed for Chapter 11 bankruptcy protection and laid off half of its workforce.

*Names changed

HISTORIC DEVELOPMENT OF TOTAL QUALITY

The Total Quality movement had its roots in the time and motion studies conducted in the early 1900s by Frederick Taylor, who is now known as the father of scientific man-agement. The timeline in Figure I–2 shows some of the major events in the evolution of the Total Quality movement since the days of Taylor.

The most fundamental aspect of scientific management was the separation of plan-ning and execution. Although this division of labor spawned tremendous leaps forward in productivity, it virtually eliminated the old concept of craftsmanship whereby one highly skilled individual performed all the tasks required to produce a quality product. In a sense, a craftsman was CEO, production worker, and quality controller all rolled into one person. Taylor's scientific management did away with this by making planning the job of management and production the job of labor. To keep quality from falling through the cracks, it was necessary to create a separate quality department. Such departments had shaky beginnings, and just who was responsible for quality became a clouded issue. (See Case Study I–4, p. 11.)

As the volume and complexity of manufacturing increased, quality became an increasingly difficult issue. The need to address increased volume and complexity gave

====== **CASE STUDY I–3** ======

Commitment and Perseverance at Hub, Inc.

Ben Camp, CEO of Hub, Inc., a metal products manufacturer outside of Atlanta, found that total quality takes commitment and perseverance. "The company studied various gurus and fashioned its own program from them. That program requires, among other things, meticulous attention to detail. It demands perseverance. It involves dismounting from managerial high horses and realizing that employees recognize lip service as soon as the mouths begin to move. Thus, getting down to action is where the easier-said-than-done part comes in. Everyone agrees that profitable companies with devoted customers and dynamic employees are good things. But not everyone wants to endure the pain of change, much less admit that, after years of effort, there still may be as much to be done as has been done. . . ."

After four years of total quality, attention to detail and perseverance had allowed Hub to identify and categorize most of its errors, a major step in correcting future performance. The categories of errors identified were defects, carrier/vendor, pricing, order entry, wrong material, wrong quality, customer error, surplus, and duplicates. In one sample month, errors in these categories cost the company over $16,000. In the same month, unidentified errors cost the company another $5,161. The errors that have been identified can be used for improving the processes that caused them. Unidentified errors cannot be corrected. Hub is committed to identifying and correcting 100 percent of its process errors. The company intends to persevere in this goal no matter how long it takes. If Hub is going to compete, it has no other option.[12]

---

birth to quality engineering in the 1920s and to reliability engineering in the 1950s. Quality engineering, in turn, gave birth to the use of statistical methods in the control of quality, which eventually led to the concept of control charts and statistical process control. Statistical process control is now a fundamental aspect of the Total Quality approach and is discussed in Chapter 16.

Joseph M. Juran, writing on the subject of quality engineering, says:

> This specialty traces its origin to the application of statistical methods for control of quality in manufacture. Much of the pioneering theoretical work was done in the 1920s by the quality assurance department of the Bell Telephone laboratories. The staff members included Shewhart, Dodge, and Edwards. Much of the pioneering application took place (also in the 1920s) within the Hawthorne Works of the Western Electric Company."[13]

Reliability engineering, which emerged in the 1950s, began a trend away from the traditional after-the-fact approach to quality control and toward inserting quality control measures throughout design and production. However, for the most part, quality control in the 1950s and 1960s involved inspections that resulted in nothing more than a cutting out of bad parts. This is illustrated by Case Study I–5 (p. 12).

Year	Milestone
1911	Frederick W. Taylor publishes *The Principles of Scientific Management,* giving birth to such techniques as time and motion studies.
1931	Walter A. Shewhart of Bell Laboratories introduces statistical quality control in his book *Economic Control of Quality of Manufactured Products.*
1940	W. Edwards Deming assists the U.S. Bureau of the Census in applying statistical sampling techniques.
1941	W. Edwards Deming joins the U.S. War Department to teach quality-control techniques.
1950	W. Edwards Deming addresses Japanese scientists, engineers, and corporate executives on the subject of quality.
1951	Joseph M. Juran publishes the *Quality Control Handbook.*
1961	Martin Company (later Martin-Marietta) builds a Pershing missile that has zero defects.
1970	Philip Crosby introduces the concept of *zero defects.*
1979	Philip Crosby publishes *Quality is Free.*
1980	Television documentary *If Japan Can . . . Why Can't We?* airs giving W. Edwards Deming renewed recognition in the U.S.
1981	Ford Motor Company invites W. Edwards Deming to speak to its top executives, which begins a rocky but productive relationship between the automaker and the quality expert.
1982	W. Edwards Deming publishes *Quality, Productivity, and Competitive Position.*
1984	Philip Crosby publishes *Quality Without Tears: The Art of Hassle-Free Management.*
1987	U.S. Congress creates the Malcolm Baldrige National Quality Award.
1988	Secretary of Defense Frank Carlucci directs the U.S. Department of Defense to adopt total quality.
1989	Florida Power and Light wins Japan's coveted Deming Prize, the first non-Japanese company to do so.
1993	The total-quality approach is widely taught in U.S. colleges and universities.

Figure I–2
Selected Historic Milestones in the Quality Movement in the United States

World War II had an impact on quality that is still being felt to this day. In general the impact was negative for the United States and positive for Japan. Because of the urgent need to meet production schedules during the war, U.S. companies focused more on meeting delivery dates than on quality. This approach became a habit that carried over into peace-time manufacturing. Japanese companies, on the other hand, in order to rebuild, were forced to learn to compete with the rest of the world in the production of non-military goods. At first their attempts were unsuccessful, and the phrase "Made in

====== **CASE STUDY I–4** ======

Early Impact of Scientific Management

"To restore the balance, the factory managers adopted a new strategy: a central inspection department headed by a chief inspector. The various departmental inspectors were then transferred to the new inspection department over the bitter opposition of the production supervisors.

"Note that during this progression of events the priority given to quality declined significantly. In addition, the responsibility for leading the quality function became vague and confused. In the days of the craft shops, the master (then also the chief executive) participated personally in the process of managing for quality. What emerged was a concept in which upper management became detached from the process of managing for quality."[14]

Japan" became synonymous with poor quality as it had been before World War II. By about 1950, however, Japan decided to get serious about producing quality products. Here is how Joseph M. Juran describes the start of the Japanese turn around:

> "To solve their quality problems the Japanese undertook to learn how other countries managed for quality. To this end the Japanese sent teams abroad to visit foreign companies and study their approach, and they translated selected foreign literature into Japanese. They also invited foreign lecturers to come to Japan and conduct training courses for managers.
>
> "From these and other inputs the Japanese devised some unprecedented strategies for creating a revolution in quality. Several of those strategies were decisive:
>
> 1. The upper managers personally took charge of leading the revolution.
> 2. All levels and functions underwent training in managing for quality.
> 3. Quality improvement was undertaken at a continuing, revolutionary pace.
> 4. The workforce was enlisted in quality improvement through the QC-concept."[15]

Case Study I–6 (p. 13) explains how Japanese manufacturers overcame a reputation for producing cheap, shabby products and developed a reputation as world leaders in the production of quality products. More than any other single factor, it was the Japanese miracle (which was not a miracle at all, but the result of a concerted effort that took twenty years to really bear fruit), that got the rest of the world to focus on quality. Once Western companies finally realized that quality was the key factor in global competition, they responded. Unfortunately, their first responses were the opposite of what was needed.

Joseph M. Juran describes those initial responses as follows: " . . . the responses to the Japanese quality revolution took many directions. Some of these directions consisted of strategies that had no relation to improving American competitiveness in quality. Rather, these were efforts to block imports through restrictive legislation and quotas, criminal prosecutions, civil lawsuits, and appeals to buy American."[16]

In spite of these early negative reactions, Western companies began to realize that the key to competing in the global marketplace was to improve quality. With this realiza-

=== **CASE STUDY I–5** ===

Early Inspection-Oriented Quality Program

"The central activity of these quality-oriented departments remained that of inspection and test—that is, separating good products from bad. The prime benefit of this activity was to reduce the risk that defective products would be shipped to customers. However, there were serious detriments: This centralized activity of the quality department helped to foster a widespread belief that achievement of quality was solely the responsibility of the quality department. In turn, this belief hampered efforts at eliminating the causes of defective products; the responsibilities were confused. As a result, failure-prone products and incapable processes remained in force and continued to generate high costs of poor quality."[17]

---

tion, the Total Quality movement finally began to gain momentum. The United States is now beginning to regain market share lost to Japan in such critical sectors as the automobile industry.

RATIONALE FOR TOTAL QUALITY

The rationale for Total Quality can be found in the need to compete in the global marketplace. Countries that are competing successfully in the global marketplace are seeing their gross national product (GNP) increase and their quality of living improve correspondingly. Countries that are not able to compete in the global marketplace are seeing their quality of life decline. Although the United States emerged from World War II as the international leader in manufacturing productivity, by the mid-1970s other countries, particularly Japan, were challenging this status. Case Study I–7 (p. 14) describes how this situation evolved.

Reversing this situation is critical. The United States built one of the highest standards of living in history by dominating world markets in the very industrial sectors that are now being challenged by foreign competitors: automobiles, computers, consumer electronics, textiles, and industrial machinery. As a result, as the United States entered the final decade of the twentieth century, its standard of living was declining rather than improving.

Starting even before the 1980s, bright, talented, and determined people have been working to reverse the erosion of U.S. market share to foreign competitors. Since this is the case, why have they not been more successful? Stephen Uselac gives the following explanation:

"American managers have traditionally been well-versed in the production, marketing, financial management information systems, and quantitative aspects of their world. However, many managers who have excellent conceptual and technical skills tend to be inadequately prepared in people skills and technical process skills. They seem to have a difficult time lead-

=============== **CASE STUDY I–6** ===============

How Japan Caught Up with the West

Immediately following World War II, the quality of products produced by Japanese companies was not good enough to compete in the international marketplace. The only advantage Japanese companies had was price. Japanese goods, as a rule, were cheap. For this reason Western manufacturers, particularly those in the United States, saw the Japanese threat as being rooted in cost rather than quality.

Reading the future more accurately, albeit belatedly, Japanese companies saw quality as the key to success and, in 1950, began doing something about it. While Japanese companies were slowly but patiently and persistently creating a quality-based infrastructure (people, processes, and facilities), Western companies, still focused on cost, were shifting the manufacture of labor-intensive products off-shore, and, at the same time, neglecting infrastructure improvements.

As a result, by the mid-1970s the quality of Japanese manufactured goods in such key areas as automobiles and consumer electronics products was often better than that of competing Western firms. As a result, Japanese exports increased exponentially while those of Western countries experienced corresponding decreases.

---

ing people to work together toward the mission of the organization. The lack of people and process skills is common in both private and public sectors. The most significant problem is our lack of commitment to the organizational, managerial, and logistical changes that must occur in order to become or remain competitive. We must design organizational systems that require coordination and cooperation between functional areas if we are to meet the competitive challenge. We must learn to work smarter, not harder. This means working together at all levels in an organization."[18]

Uselac makes an important point here. The areas in which U.S. managers historically have excelled—finance, management information systems, and the other quantitative aspects of doing business—are important, but they have no direct bearing on the quality of a product or on the people and processes involved in designing, producing, and testing it.

Uselac attributes the situation to a lack of managers who have strong people and process skills, which is true. But why is this the case? Is the United States incapable of producing such managers, or is there another reason? There is, in fact, another reason: emphasis. The United States has excelled in the quantitative aspects of the job because these are the aspects that historically have been emphasized, rewarded, and given status. In other words, over time the glamour jobs in U.S. industry have become support jobs that have little to do with product quality. Evidence of this misplaced emphasis can be seen in how U.S. industry historically has spent its training dollars, where it has offered the most rewarding financial incentives, and even in which major programs of study have been emphasized by the nation's colleges and universities. During the critical

Why Total Quality Is Important in the United States

The United States has enjoyed one of the highest standards of living in the world for well over 100 years. We have built the largest and best schools, hospitals, communities, factories, and shopping services worldwide. We have been a leader in productivity growth and innovation, which are primary stimuli for raising the standard of living. Our manufacturing capabilities have given this nation an economic base that has permitted us to build a society that has set the standard for excellence around the globe. However, there is increasing concern about the future of manufacturing in this country. Since 1980, the United States has experienced a loss of markets, lagging productivity, high unemployment in the manufacturing sector, and an eroding competitive position in the world. All of these are well-documented symptoms of American industry's decline. Foreign competitors, particularly the Japanese, have captured many markets in which American industry was formerly dominant.

It is clear that we have lost our predominant position in a number of key industries, such as steel, automotive, industrial machinery, textiles, and electronics. Furthermore, it is apparent that competitive pressures continue to mount. Foreign competition offering high quality, low-priced products continues to edge into our marketplace.[19]

decade of the 1980s, which saw the U.S. experience its most significant declines in market share, the college degree of choice among American students in business schools nationwide was the MBA. During the same decade, U.S. colleges of engineering enrolled significantly more foreign-born students than U.S. students. The problem of misplaced emphasis can be summarized as follows: While competing countries have been improving their people and process skills, the United States has focused on improving its marketing and accounting skills. Said another way, if global competition in the industrial sector were an automobile race, the United States in the 1980s focused on getting more and better advertisements painted on its car while its foreign competitors emphasized improving the car, the driver, the mechanics, and the pit crews.

When U.S. firms belatedly realized that winning in the global marketplace had more to do with quality than with marketing, the Total Quality movement picked up steam. The rationale, then, is simple: The best way to win in global competition is with quality at low cost. The best way to produce quality at a low cost is to continually improve people, processes, and environments. The best way to continually improve people, processes, and environments is the Total Quality way.

KEY ELEMENTS OF TOTAL QUALITY

The Total Quality approach was defined in Figure I–1. This definition has two components: the *what* and the *how* of Total Quality. What distinguishes Total Quality from other

---

TOTAL QUALITY TIP

Quality and Competitiveness

"Quality is about doing things right the first time and about satisfying customers. But quality is also about costs, revenues, and profits. Quality plays a key role in keeping costs low, revenues high, and profits robust."[20]

Perry L. Johnson

---

approaches to doing business is the *how* component of the definition. This component has ten critical elements, each of which is explained in the remainder of this section.

Customer Focus

The customer is the driver in a Total Quality setting. This applies to both internal and external customers. External customers define the quality of the product or service delivered. Internal customers help define the quality of the people, processes, and environments associated with the products or services.

Quality expert Peter R. Scholtes explains the concept of customer focus as follows:

> Whereas Management by Results begins with profit and loss and return on investment, Quality Leadership starts with the customer. Under Quality Leadership, an organization's goal is to meet and exceed customer needs, to give lasting value to the customer. The return will follow as customers boast of the company's quality and service. Members of a quality organization recognize both external customers, those who purchase or use the products or services, and internal customers, fellow employees whose work depends on the work that precedes them."[21]

Obsession with Quality

In a Total Quality organization, internal and external customers define quality. With quality defined, the organization must then become obsessed with meeting or exceeding this definition. This means all personnel at all levels approach all aspects of the job from the perspective of "how can we do this better?" When an organization is obsessed with quality, good enough is never good enough.

Scientific Approach

Detractors of the Total Quality concept sometimes see it as nothing more than "mushy, people stuff."[22] While it is true that people skills, involvement, and empowerment are important in a Total Quality setting, they represent only a part of the equation. Another important part of the equation is the use of the scientific approach in structuring work

---

TOTAL QUALITY TIP

Quality and Customers

"In fact, customer satisfaction is regarded as the only relevant objective for ensuring stable and continuously increasing business."[23]

Giorgo Merli

---

and in decision-making and problem-solving relating to that work. This means that hard data are used in establishing benchmarks, monitoring performance, and making improvements.

Long-Term Commitment

Organizations that have a history of implementing management innovations after attending short-term seminars often fail in their initial attempt to adopt the Total Quality approach. This is because they approach Total Quality as just another management innovation rather than a whole new way of doing business that requires a whole new corporate culture.

Too few organizations begin the implementation of Total Quality with the long-term commitment to change that is necessary for success. Quality consultant Jim Clemmer of Toronto-based Achieve International describes problems that organizations starting quality initiatives frequently have, the first of which is as follows:

> Senior managers decide they want all of the benefits of Total Quality, so they hire an expert or throw some money at a particular department. Why that approach doesn't work has been widely discussed; I won't belabor the point.[24]

Teamwork

In traditionally managed organizations, the best competitive efforts are often between departments within the organization. Internal competition tends to use up energy that could and should be focused on improving quality, and, in turn, external competitiveness. Peter R. Scholtes describes the need for teamwork as follows:

> Where once there may have been barriers, rivalries, and distrust, the quality company fosters teamwork and partnerships with the workforce and their representatives. This partnership is not a pretense, a new look to an old battle. It is a common struggle for the customers, not separate struggles for power. The nature of a common struggle for quality also applies to relationships with suppliers, regulating agencies, and local communities.[25]

TOTAL QUALITY TIP

Continually Improving Processes and Systems

"Quality Leadership recognizes—as Dr. Joseph M. Juran and Dr. W. Edwards Deming have maintained since the early 1950s—that at least 85% of an organization's failures are the fault of management controlled systems. Workers can control fewer that 15% of the problems. In Quality Leadership, the focus is on constant and rigorous improvement of every system, not on blaming individuals for problems."[26]

Peter R. Scholtes

Continual Improvement of Systems

Products are developed and services delivered by people using processes within environments (systems). In order to continually improve the quality of products or services—which is a fundamental goal in a Total Quality setting—it is necessary to continually improve systems.

Education and Training

Education and training are fundamental to Total Quality because they represent the best way to improve people on a continual basis. According to Scholtes,

> In a quality organization everyone is constantly learning. Management encourages employees to constantly elevate their level of technical skill and professional expertise. People gain an ever-greater mastery of their jobs and learn to broaden their capabilities.[27]

It is through education and training that people who know how to work hard learn how to also work smart.

Freedom through Control

Involving and empowering employees is fundamental to Total Quality as a way to simultaneously bring more minds to bear on the decision-making process and increase the ownership employees feel in decisions that are made. Detractors of Total Quality sometimes mistakenly see employee involvement as a loss of management control, when in fact control is fundamental to Total Quality. The freedoms enjoyed in a Total Quality setting are actually the result of well-planned and well-carried-out controls. Scholtes explains this paradox as follows:

In Quality Leadership there is control, yet there is freedom. There is control over the best-known method for any given process. Employees standardize processes and find ways to ensure everyone follows the standard procedures. They reduce variation in output by reducing variation in the way work is done. As these changes take hold, they are freer to spend time eliminating problems, to discover new markets, and to gain greater mastery over processes.[28]

Unity of Purpose

Historically, management and labor have had an adversarial relationship in U.S. industry. One could debate the reasons behind management-labor discord *ad infinitum* without achieving consensus. Such is not the purpose here. Suffice it to say that there is probably sufficient wrong and right on both sides of the issue. From the perspective of Total Quality, who or what is to blame for adversarial management-labor relations is irrelevant. What is important is that in order to apply the Total Quality approach, organizations must have unity of purpose.

A question frequently asked about this element of Total Quality is, "Does unity of purpose mean that unions will no longer be needed?" The answer to this question is that unity of purpose has nothing to do with whether or not unions are needed. Collective bargaining is about wages, benefits, and working conditions, not corporate purpose and vision. While employees should feel more involved and empowered in a Total Quality setting than in a traditionally managed situation, the goal is to enhance competitiveness, not eliminate unions. For example, in Japan where companies are known for achieving unity of purpose, unions are still very much in evidence. Unity of purpose does not necessarily mean that labor and management will always agree on wages, benefits, and working conditions.

Employee Involvement and Empowerment

This is one of the most misunderstood elements of the Total Quality approach and one of the most misrepresented by its detractors. The basis for involving employees is twofold. First, it increases the likelihood of a good decision, a better plan, or a more effective improvement by bringing more minds to bear on the situation—and not just any minds, but the minds of the people who are closest to the work in question. Second, it promotes ownership of decisions by involving the people who will have to implement them.

Empowerment means not just involving people, but involving them in ways that give them a real voice. One of the ways this can be done is by structuring work that allows employees to make decisions concerning the improvement of work processes within well-specified parameters. Should a machinist be allowed to unilaterally drop a vendor if the vendor delivers substandard material? No. However, the machinist should have an avenue for offering her input into the matter. Should the same machinist be allowed to change the way she sets up her machine? Yes, if by doing so she can improve her part of the process without adversely affecting someone else's. Having done so, the next step should be to show other machinists her innovation so that they might try it and so that this better way she has discovered can be standardized.

TOTAL QUALITY TIP

Total Quality Requires Unity of Purpose

"There is a unity of purpose throughout the company in accord with clear and widely understood vision. This environment nurtures total commitment from all employees. Rewards go beyond benefits and salaries to the belief 'we are family' and 'we do excellent work.'"[29]

Peter R. Scholtes

TOTAL QUALITY PIONEERS

Total Quality is not just a single concept, but a number of related concepts which together create a comprehensive and different approach to doing business. Many people contributed in meaningful ways to the development of the various concepts that are known collectively as Total Quality. Of these, three who have played major roles are W. Edwards Deming, Joseph M. Juran, and Philip B. Crosby. To these three many would add Armand V. Feigenbaum and a number of Japanese experts such as Shigeo Shingo and Taiichi Ohno.

Deming's Contributions

Of the various quality pioneers in the United States, the best known is Dr. W. Edwards Deming. According to Deming biographer Andrea Gabor,

> Deming also has become by far the most influential proponent of quality management in the United States. While both Joseph Juran and Armand V. Feigenbaum have strong reputations and advocate approaches to quality that in many cases overlap with Deming's ideas, neither has achieved the stature of Deming. One reason is that while these experts have often taken very nuts-and-bolts, practical approaches to quality improvement, Deming has played the role of visionary, distilling disparate management ideas into a compelling new philosophy.[30]

Deming came a long way to achieve this status of internationally acclaimed quality expert. He comes from a very humble background. During his formative years, Deming's family bounced from small town to small town in Iowa and Wyoming, trying in vain to rise up out of poverty. These early circumstances gave Deming a life-long appreciation for economy and thrift. In later years, even after his international consulting firm was generating a substantial income, Deming conducted business out of a simple office in the basement of his modest home.

Deming worked as a janitor and at other odd jobs while earning his baccalaureate degree in engineering from the University of Wyoming. He later received a master's degree in mathematics and physics from the University of Colorado and a doctorate in physics from Yale.

Deming's only full-time employment for a corporation was with Western Electric. Many feel that what he witnessed during his employment at Western Electric had a major impact on the direction the rest of his life would take. Deming was disturbed by the amount of waste he saw at Western Electric's Hawthorne Plant. It was here that he pioneered the use of statistics in quality.

Although Deming was asked in 1940 to help the U.S. Bureau of the Census adopt statistical sampling techniques, his reception in the United States during these early years was not positive. With little real competition in the international marketplace, major U.S. corporations felt no need of his help, nor did corporations from other countries. However, the advent of World War II changed all this, and Deming was soon on the road to becoming, in the words of biographer Andrea Gabor, "the man who discovered quality."[31]

World War II put almost all of Japan's industry in the business of producing war materials. After the war, Japanese firms had to convert to the production of consumer goods, and the conversion did not go well. In order to have a market for their products, Japanese firms had to enter the international marketplace in direct competition with companies from the other industrialized countries. The Japanese firms did not fare well, and Japanese products became synonymous with terms such as "cheap" and "junk."

By the late 1940s, key industrial leaders in Japan had finally come to the realization that quality was the key to competing in the international marketplace. At this time, Shigeiti Mariguti of Tokyo University, Sizaturo Mishibori of Toshiba, and several other Japanese leaders invited Deming to visit Japan and share his views on quality. Unlike their counterparts in the United States, Japanese industrialists accepted Deming's views, learned his techniques, and adopted his philosophy. So powerful was Deming's impact on industry in Japan that the most coveted award a Japanese company can win is called the Deming Prize. In fact, the standards that must be met in order to win this prize are so difficult and so strenuously applied that some Japanese companies are now questioning its high price (see Case Study I–8).

By the 1980s, leading industrialists in the United States were where their Japanese counterparts had been in the late 1940s. At long last, Deming's services began to be requested in his own country. By this time Deming was over 80 years old. He has not been received as openly and warmly in the United States as he was in Japan. As a result, Deming's attitude toward corporate executives in the United States was cantankerous at best. Andrea Gabor gives the following example of Deming's dealings with executives from Ford Motor Company:

> The initial contacts were unsettling for Ford. Instead of delivering a slick presentation on how the automaker could solve its quality problems—the sort of thing that became the stock in trade of U.S. quality experts during the 1980s, Deming questioned, rambled, and seemed to take pleasure in making a laughingstock of his listeners. During the first meeting, wearing one of his signature timeworn three-piece suits, Deming glowered at the car executives with steely blue eyes.[32]

It would be difficult to overstate Deming's contributions to the Total Quality movement. Many consider him the father of the movement. Specifically, the things he is most

===== **CASE STUDY I–8** =================================

Japanese Firms Question the Deming Prize

"Some executives contend that even competing for the Deming Prize can sometimes be counter-productive. For example, Shimizu Construction, which won the Deming Prize in 1983, spent so much time and effort chasing the prize that same year its financial results showed a clear drop from the year before. The human toll can also be high. Some companies go so far as to gather their middle management at the foot of Ft. Fuji for a kind of hazing session, subjecting the group to seven days of marathon discussions that last from 5:00 a.m. to 10:00 p.m., during which each participant admits he has been neglecting his work. Sachiaki Nagae of TIT says that for the last year before going for the Deming Prize he and his managers went without vacations and routinely worked seven days a week."[33]

---

widely known for are the Deming Cycle, his Fourteen Points, and the Seven Deadly Diseases. The Deming Cycle is summarized in Figure I–3. The Deming Cycle was developed to link the production of a product with consumer needs and to focus the resources of all departments—research, design, production, marketing—in a cooperative effort to meet those needs. The Deming Cycle proceeds as follows:

1. Conduct consumer research and use it in planning the product (plan).
2. Produce the product (do).
3. Check the product to make sure it was produced in accordance with the plan (check).
4. Market the product (act).
5. Analyze how the product is received in the marketplace in terms of quality, cost, and other criteria.

Deming's Fourteen Points

Deming's philosophy is both summarized and operationalized by his Fourteen Points. Peter R. Scholtes describes Deming's Fourteen Points as follows:

> Over the years, Dr. Deming has developed 14 points that describe what is necessary for a business to survive and be competitive today. At first encounter, their meaning may not be clear. But they are the very heart of Dr. Deming's philosophy. They contain the essence of all his teachings. Read them, think about them, talk about them with your co-workers or with experts who deeply understand the concepts. And then come back to think about them again. Soon you will start to understand how they work together and their significance in the true quality organization. Understanding the 14 points can shape a new attitude toward work and the work environment that will foster continuous improvement.[34]

Deming's Fourteen Points are contained in Figure I–4. The reader should keep in mind that Dr. Deming has modified the specific wording of various points over the years. This accounts for the minor differences that may be noticed when the Fourteen Points

Figure I–3
The Deming Cycle

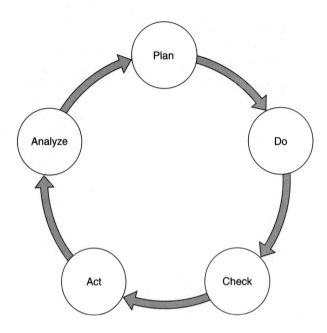

1. Create constancy of purpose toward the improvement of products and services in order to become competitive, stay in business, and provide jobs.
2. Adopt the new philosophy. Management must learn that it is a new economic age and awaken to the challenge, learn their responsibilities, and take on leadership for change.
3. Stop depending on inspection to achieve quality. Build in quality from the start.
4. Stop awarding contracts on the basis of low bids.
5. Improve continuously and forever the system of production and service, to improve quality and productivity, and thus constantly reduce costs.
6. Institute training on the job.
7. Institute leadership. The purpose of leadership should be to help people and technology work better.
8. Drive out fear so that everyone may work effectively.
9. Break down barriers between departments so that people can work as a team.
10. Eliminate slogans, exhortations, and targets for the workforce. They create adversarial relationships.
11. Eliminate quotas and management by objectives. Substitute leadership.
12. Remove barriers that rob employees of their pride of workmanship.
13. Institute a vigorous program of education and self-improvement.
14. Make the transformation everyone's job and put everyone to work on it.

Figure I–4
Deming's Fourteen Points

1. Lack of constancy of purpose to plan products and services that have a market sufficient to keep the company in business and provide jobs.
2. Emphasis on short-term profits; short-term thinking that is driven by a fear of unfriendly takeover attempts and pressure from bankers and shareholders to produce dividends.
3. Personal review systems for managers and management by objectives without providing methods or resources to accomplish objectives. Performance evaluations, merit ratings, and annual appraisals are all part of this *disease*.
4. Job hopping by managers.
5. Using only visible data and information in decision-making with little or no consideration given to what is not known or cannot be known.
6. Excessive medical costs.
7. Excessive costs of liability driven up by lawyers that work on contingency fees.

Figure I–5
Deming's Seven Deadly Diseases

are seen in different publications. Another point is that Dr. Deming has stated repeatedly in his later years that if he had it all to do over again, he would leave off the numbers. Many people, in Deming's opinion, interpret numbers as an order of priority or progression, when in fact this is not the case. The numbers represent neither an order of progression nor relative priorities.[35]

Deming's Seven Deadly Diseases

The Fourteen Points summarize Deming's views on what a company must do to effect a positive transition from business as usual to world-class quality. His Seven Deadly Diseases (listed in Figure I–5) summarize Deming's views on factors that can inhibit such a transformation.

These factors ring particularly true when viewed from the perspective of U.S. firms trying to compete in the global marketplace. While some of these factors can be eliminated by adopting Total Quality, three of them cannot, which does not bode well for U.S. firms trying to regain market share. Total Quality can either eliminate or reduce the impact of a lack of consistency, personal review systems, job hopping, and using only visible data. However, Total Quality will not free corporate executives from pressure to produce short-term profits, excessive medical costs, or excessive liability costs. These are diseases of the nation's financial, health care, and legal systems. By finding ways for business and government to cooperate appropriately without collaborating inappropriately, other industrialized countries have been able to focus their industry on long-term rather than short-term profits, hold down health care costs, and prevent the proliferation of costly litigation that has occurred in the United States. Excessive health care and legal costs represent non-value-added costs that must be added to the cost of products produced and services delivered in the United States.

Research and Development
We dedicate ourselves to identifying proven roads that lead to enviable results, and to creating products and services that reflect the experience of successful quality efforts.

Consulting Services
Our quality professionals counsel and assist managers in assessing needs, developing strategies, and applying specific methods for achieving quality goals.

Seminars and Workshops
Juran courses are comprehensive and practical, addressing a full spectrum of needs, industries, and functions. Courses are presented worldwide at client organizations and as public offerings.

Training and Support Materials
Juran videos, workbooks, texts, and reference books are more widely used and translated than those of any other source.

Figure I–6
Quality-Related Services Provided Worldwide by the Juran Institute, Inc.
From promotional literature of Juran Institute, Inc., 1992.

Juran's Contributions

Joseph M. Juran ranks near Deming in the contributions he has made to quality and the recognition he has received as a result. The Juran Institute, Inc. in Wilton, Connecticut is an international leader in conducting training, research, and consulting activities in the area of quality management (see Figure I–6). Quality-related materials produced by Juran have been translated into 14 different languages.

Juran holds degrees in both engineering and law. The Emperor of Japan awarded him the Order of the Sacred Treasure medal in recognition of his efforts to promote quality in Japan and for encouraging friendship between Japan and the United States. Juran is best known for the following contributions to the quality philosophy: Juran's

1. Achieve structured improvements on a continual basis combined with dedication and a sense of urgency.
2. Establish an extensive training program.
3. Establish commitment and leadership on the part of higher management.

Figure I–7
Juran's Three Basic Steps to Progress
Stephen Uselac, *Zen Leadership: The Human Side of Total Quality Team Management* (Londonville, OH: Mohican Publishing Company, 1993), 37.

TOTAL QUALITY TIP

Quality and the Law of Diminishing Returns

"Juran favors the concept of quality circles because they improve communication between management and labor. Furthermore, he recommends the use of statistical process control, but does believe that quality is not free. He explains that within the law of diminishing returns, quality will optimize, and beyond that point conformance is more costly than the value of the quality obtained."[36]

Three Basic Steps to Progress, Juran's Ten Steps to Quality Improvement, the Pareto Principle, and the Juran Trilogy.

Juran's Three Basic Steps to Progress are listed in Figure I–7. These three broad steps are, in Juran's opinion, steps that companies must take in order to achieve world-class quality. He also believes there is a point of diminishing returns that applies to quality and competitiveness. An example illustrates his point. Say that an automobile maker's research on its mid-range line of cars reveals that buyers drive them an average of 50,000 miles before trading them in. Applying Juran's theory, this automaker should invest the resources necessary to make this line of cars run trouble-free for perhaps 60,000 miles. According to Juran, resources devoted to improving quality beyond this point will run the cost up higher than the typical buyer is willing to pay.

Juran's Ten Steps to Quality Improvement are listed in Figure I–8. Upon examining these steps, you will note some overlap between them and Deming's Fourteen Points.

1. Build awareness of both the need for improvement and opportunities for improvement.
2. Set goals for improvement.
3. Organize to meet the goals that have been set.
4. Provide training.
5. Implement projects aimed at solving problems.
6. Report progress.
7. Give recognition.
8. Communicate results.
9. Keep score.
10. Maintain momentum by building improvement into the company's regular systems.

Figure I–8
Juran's Ten Steps to Quality Improvement
Stephen Uselac, *Zen Leadership: The Human Side of Total Quality Team Management* (Londonville, OH: Mohican Publishing Company, 1993), 37.

TOTAL QUALITY TIP

The Pareto Principle

"This principle is sometimes called the ⁸⁰⁄₂₀ rule: 80% of the trouble comes from 20% of the problems. Though named for turn-of-the-century economist Vilfredo Pareto, it was Dr. Juran who applied the idea to management. Dr Juran advises us to concentrate on the 'vital few' sources of problems and not be distracted by those of lesser importance."[37]

Juran's ten steps also mesh well with the philosophy of quality experts whose contributions are explained later in this chapter.

The Pareto Principle

The Pareto Principle—that 80 percent of the trouble comes from 20 percent of the problems—espoused by Juran shows up in the views of most quality experts, although often by other names. According to this principle, organizations should concentrate their energy on eliminating the vital few sources that cause the majority of problems. Further, both Juran and Deming believe that systems that are controlled by management are the systems in which the majority of problems occur.

The Juran Trilogy

The Juran Trilogy, illustrated in Figure I–9, summarizes the three primary managerial functions. Juran's views on these functions are explained as follows:[38]

Quality Planning.   Quality planning involves developing the products, systems, and processes needed to meet or exceed customer expectations. The following steps are required:

1. Determine who the customers are.
2. Identify the needs of customers.
3. Develop products with features that respond to customer needs.
4. Develop systems and processes that allow the organization to produce these features.
5. Deploy the plans to operational levels.

Quality Control.   The control of quality involves the following steps:

1. Assess actual quality performance.
2. Compare performance with goals.
3. Act on differences between performance and goals.

Figure I–9
The Juran Trilogy*
*The Juran Trilogy® is a registered trademark of the Juran Institute, Inc.

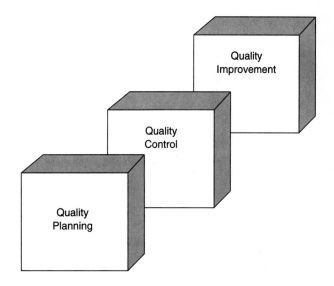

Quality Improvement.   The improvement of quality should be ongoing and continual. Quality improvement involves the following steps:

1. Develop the infrastructure necessary to make annual quality improvements.
2. Identify specific areas in need of improvement and implement improvement projects.
3. Establish a project team with responsibility for completing each improvement project.
4. Provide teams with what they need in order to be able to diagnose problems to determine root causes, develop solutions, and establish controls that will maintain gains made.

Crosby's Contributions

Philip B. Crosby's corporate background includes 14 years as Director of Quality at ITT (1965–1979). He left ITT in 1979 to form Philip Crosby Associates, an international consulting firm on quality improvement, which he ran until 1992. He then retired as CEO to devote his time to lecturing on quality-related issues.

Crosby, who defines quality simply as conformance, is best known for his advocacy of zero-defects management and error prevention as opposed to pursuit of statistically acceptable levels of quality. He is also known for his fourteen steps to quality improvement (listed in Figure I–10), and for his "quality vaccine," which consists of three ingredients: determination, education, and implementation.[39]

WHY TOTAL QUALITY SOMETIMES FAILS

When organizations approach Total Quality as just another management innovation or, even worse, as a quick fix, their efforts are doomed to fail from the start. Unfortunately,

1. Make it clear that management is committed to quality for the long term.
2. Form cross-departmental quality teams.
3. Identify where current and potential problems exist.
4. Assess the cost of quality and explain how it is used as a management tool.
5. Increase the quality awareness and personal commitment of all employees.
6. Take immediate action to correct problems identified.
7. Establish a zero defects program.
8. Train supervisors to carry out their responsibilities in the quality program.
9. Hold a Zero Defects Day to ensure all employees are aware there is a new direction.
10. Encourage individuals and teams to establish both personal and team improvement goals.
11. Encourage employees to tell management about obstacles they face in trying to meet quality goals.
12. Recognize employees who participate.
13. Implement quality councils to promote continual communication.
14. Repeat everything to illustrate that quality improvement is a never-ending process.

Figure I–10
Crosby's Fourteen Steps to Quality Improvement
Stephen Uselac, *Zen Leadership: The Human Side of Total Quality Team Management* (Londonville, OH: Mohican Publishing Company, 1993), 37.

there are organizations that have taken this approach and, as a result of their inevitable failure, have become detractors of Total Quality.

Other organizations have approached Total Quality with the necessary commitment, but with unrealistic expectations. One such example is Douglas Aircraft, a subsidiary of McDonnell-Douglas Corporation, which implemented Total Quality in an attempt to become less susceptible to economic downturns in the volatile aircraft business. The company's unionized labor force embraced Total Quality as a way to prevent layoffs. When a depressed economy—something Total Quality cannot control—forced Douglas Aircraft to make massive layoffs, labor-management relations soured to the point that Total Quality became impossible.

Common Errors Made When Starting Quality Initiatives

In addition to making half-hearted implementation efforts and having unrealistic expectations, organizations commonly make several other errors when starting quality initiatives. Jim Clemmer of Achieve International has identified the following five commonly made errors:[40]

TOTAL QUALITY TIP

Total Quality Has Believers and Detractors

"Only yesterday, total quality management, or TQM, was on everyone's lips as the miracle that could save America's businesses from liquidation, its workforce from terminal disgruntlement and its industries from global obsolescence. Today the movement is being assailed by some as faddish, more characterized by style than sincerity, and even destructive—promising employees profound changes that never materialize. Despite the grumbling, the movement still has plenty of support. People willing to understand and stick with it describe it as a continuous process in which an organization statistically analyzes how jobs are done; disposes of procedures that don't work; uses all employees' broadest expertise and rewards it meaningfully; purges itself of stereotypical thinking."[41]

Atlanta Constitution

Senior Management Delegation and Poor Leadership   Some organizations attempt to start a quality initiative by delegating responsibility to a hired expert rather than applying the leadership necessary to get everyone involved.

Team Mania   Ultimately, teams should be established and all employees should be involved with them. However, working in teams is an approach that must be learned. Supervisors must learn how to be effective coaches and employees must learn how to be team players. The organization must undergo a cultural change before teamwork can succeed. Rushing in and putting everyone in teams before learning has occurred and the corporate culture has changed will create problems rather than solve them.

Deployment Process   Some organizations develop quality initiatives without concurrently developing plans for integrating them into all elements of the organization (i.e., operations, budgeting, marketing, etc.). According to Clemmer, "More time must be spent preparing plans and getting key stakeholders on board, including managers, unions, suppliers, and other production people. It takes time to pull them in. It involves thinking about structure, recognition, skill development, education, and awareness."[42]

Taking a Narrow, Dogmatic Approach   Some organizations are determined to take the Deming approach, Juran approach, or Crosby approach and use only the principles prescribed therein. None of the approaches advocated by these and other leading quality experts is truly a one-size-fits-all proposition. Even the experts encourage organizations to tailor quality programs to their individual needs.

Confusion about the Differences among Education, Awareness, Inspiration, and Skill-Building   According to Clemmer, "You can send people to five days of training in group dynamics, inspire them, teach them managerial styles, and show them all sorts of grids and analysis, but that doesn't mean you've built any skills. There is a time to educate and inspire and make people aware, and there is a time to give them practical tools they can use to do something specific and different than they did last week."[43]

IMPLEMENTING TOTAL QUALITY

This introduction has provided an overview of the what, why, when, and where of Total Quality. For a more in-depth treatment of these conceptual aspects, refer to *Introduction to Total Quality* by Goetsch and Davis.[44] The remainder of this book is devoted to a comprehensive treatment of the how or implementing aspects of Total Quality. This section provides an overview of the implementation process.

Implementation Steps

Figure I–11 is a graphic representation of the steps involved in implementing Total Quality and the people responsible for them. Notice from this figure that the implementation steps are divided into three phases: preparation, planning, and execution. These phases are described in the following paragraphs:

Preparation Phase

The preparation phase encompasses Steps 1–11. Responsibility for carrying out the activities associated with these steps belongs to the organization's top executive, the Total Quality consultant or in-house expert, the Total Quality Steering Committee, and the Steering Committee augmented by additional members. The various steps in this phase are explained in Steps 1–11 of this book.

Planning Phase

The planning phase encompasses Steps 12–16. Responsibility for carrying out the activities associated with this phase belongs to the Total Quality Steering Committee. The various steps in this phase are explained in Steps 12–16.

Execution Phase

The execution phase encompasses Steps 17–20. Responsibility for carrying out the activities associated with these steps belongs to the project teams and the Steering Committee. The various steps in this phase are explained in Steps 17–20.

SUMMARY

1. Quality has been defined in a number of different ways. When viewed from a consumer's perspective, it means meeting or exceeding customer expectations.
2. Total Quality is an approach to doing business that attempts to maximize the competitiveness of an organization through the continual improvement of the quality of its products, services, people, processes, and environments.

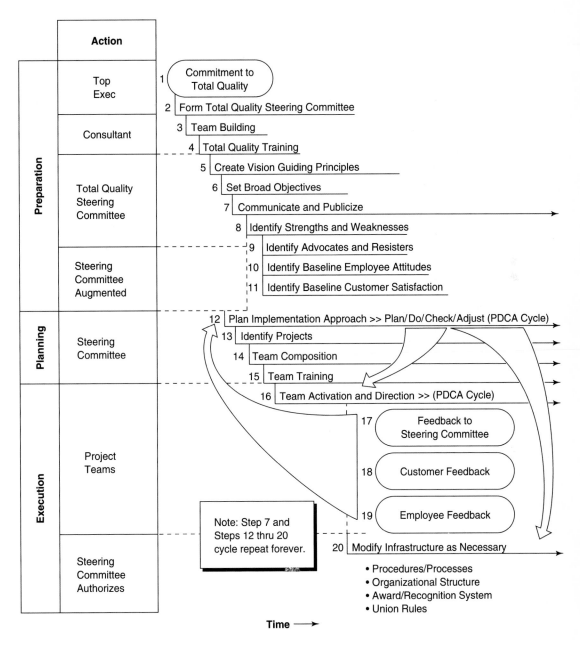

Action

Preparation		
Top Exec	1	Commitment to Total Quality
	2	Form Total Quality Steering Committee
Consultant	3	Team Building
	4	Total Quality Training
Total Quality Steering Committee	5	Create Vision Guiding Principles
	6	Set Broad Objectives
	7	Communicate and Publicize
	8	Identify Strengths and Weaknesses
Steering Committee Augmented	9	Identify Advocates and Resisters
	10	Identify Baseline Employee Attitudes
	11	Identify Baseline Customer Satisfaction

Planning

Steering Committee

12 Plan Implementation Approach >> Plan/Do/Check/Adjust (PDCA Cycle)
13 Identify Projects
14 Team Composition
15 Team Training
16 Team Activation and Direction >> (PDCA Cycle)

Execution

Project Teams

17 Feedback to Steering Committee
18 Customer Feedback
19 Employee Feedback

Steering Committee Authorizes

Note: Step 7 and Steps 12 thru 20 cycle repeat forever.

20 Modify Infrastructure as Necessary
• Procedures/Processes
• Organizational Structure
• Award/Recognition System
• Union Rules

Time ⟶

Figure I–11
Implementation Steps for Total Quality

31

3. The phrase *Total Quality* is used rather than a three-letter acronym to keep the concept from being perceived as just another management gimmick.
4. Key characteristics of the Total Quality approach are as follows: customer focus, an obsession with quality, a scientific approach, long-term commitment, teamwork, employee involvement and empowerment, continual process improvement, bottom-up education and training beginning with line employees, freedom through control, and unity of purpose.
5. The rationale for Total Quality can be found in the need to compete in the global marketplace. Countries that are competing successfully in the global marketplace are seeing their standard of living improve. Those that are not competing successfully are seeing their standard of living decline.
6. W. Edwards Deming is best known for his Fourteen Points, the Deming Cycle, and the Seven Deadly Diseases.
7. Joseph M. Juran is best known for Juran's Three Basic Steps to Progress, Juran's Ten Steps to Quality Improvement, the Pareto Principle, and the Juran Trilogy.
8. Philip B. Crosby is best known for his quality vaccine, which comprises determination, education, and implementation, and for his Fourteen Steps to Quality Improvement.
9. Common errors made when starting quality initiatives include senior management delegation and poor leadership; team mania; the deployment process; taking a narrow, dogmatic approach; and confusion about the differences among education, awareness, inspiration, and skill-building.

KEY TERMS AND CONCEPTS

Baseline customer satisfaction	Long-term commitment
Baseline employee attitudes	Obsession with quality
Bottom-up education and training	The Pareto Principle
Communicate and publicize	PDCA Cycle
Continual process improvement	Planning
Customer feedback	Preparation
Customer focus	Quality
Deming Cycle	Quality control
Deming's Fourteen Points	Quality improvement
Deming's Seven Deadly Diseases	Quality planning
Employee feedback	Quality vaccine
Employee involvement and empowerment	Scientific approach
Execution	Steering Committee
Freedom through control	Team composition
The Juran Trilogy	Team training

Teamwork

Total Quality

Total Quality Control (TQC)

Total Quality Leadership (TQL)

Total Quality Management (TQM)

Unity of purpose

REVIEW QUESTIONS

1. Define the term quality.
2. What is Total Quality?
3. Why do the authors of this book use the term *Total Quality* instead of an acronym?
4. List and explain the key elements of Total Quality.
5. Explain the rationale for the Total Quality approach to doing business.
6. Describe the following concepts:
   - Deming's Fourteen Points
   - The Deming Cycle
   - The Seven Deadly Diseases
7. List and explain Juran's main contributions to the quality movement.
8. Why do some quality initiatives fail?
9. For what contributions to the quality movement is Philip B. Crosby known?
10. Summarize the most common errors made when starting quality initiatives.
11. Summarize the 20 steps for implementing Total Quality.

ENDNOTES

1. Stephen Uselac, *Zen Leadership: The Human Side of Total Quality Team Management* (Londonville, OH: Mohican Publishing Company, 1993), 20.
2. Air Force Development Test Center, *Total Quality Management (TQM) Training Package,* 1991, 8.
3. Jerry Romano, "It's Time for a Quality Management Revolution." Workshop presented to the Emerald Coast Personnel Manager's Association on May 20, 1992.
4. Romano, 1.
5. Romano, 1.
6. W. Edwards Deming, *Out of the Crisis* (Cambridge, MA: Massachusetts Institute of Technology Center for Advanced Engineering Study, 1986), 168.
7. Deming, 169.
8. Perry L. Johnson, *Total Quality Management* (Southfield, MI: Perry Johnson, Inc., 1991), 1-1.
9. Deming, 169.
10. Irwin Bross, *Design for Decision* (New York: Macmillan Publishing Company, 1953), 95. As quoted by W. Edwards Deming, *Out of the Crisis* (Cambridge, MA: Massachusetts Institute of Technology Center for Advanced Engineering Study. 1986), 168.
11. Air Force Development Test Center, 13.

12. "Total Quality: Management Method is More than Words; Time and Commitment are Required," *Atlanta Constitution*, October 11, 1992, R-1 and R-6.

13. J. M. Juran, *Juran on Leadership for Quality.* (New York: The Free Press, 1989), 5.

14. Juran, 4.

15. Juran, 7–8.

16. Juran, 10.

17. Juran, 6.

18. Uselac, 4.

19. Uselac, 3.

20. Johnson, 9-1.

21. Peter R. Scholtes, *The Team Handbook* (Madison, WI: Joiner Associates, Inc., 1992), 1-11.

22. Comment made by a participant in a workshop on Total Quality presented by the author in Fort Walton Beach, Florida, in January 1993.

23. Giorgo Merli, *Total Manufacturing Management* (Cambridge, MA: Productivity Press, 1990), 84.

24. Jim Clemmer, "Eye on Quality," *Total Quality Newsletter,* Volume 3, Number 4, April 1992, 7.

25. Scholtes, 1-13.

26. Scholtes, 1–12.

27. Scholtes, 1–13.

28. Scholtes, 1–12.

29. Scholtes, 1–12.

30. Andrea Gabor, *The Man Who Discovered Quality* (New York: Times Books, Random House, 1990), 16.

31. Gabor, 1.

32. Gabor, 136.

33. Gabor, 97.

34. Scholtes, 2–4.

35. W. Edwards Deming, from comments made during a teleconference on Total Quality broadcast by George Washington University in January 1992.

36. Uselac, 37.

37. Scholtes, 2–9.

38. Juran, 20.

39. Uselac, 38.

40. Jim Clemmer, "5 Common Errors Companies Make Starting Quality Initiatives," *Total Quality*, April 1992, Volume 3, Number 4, 7.

41. "Total Quality," R-1.

42. Clemmer, 7.

43. Clemmer, 7.

44. David L. Goetsch & Stanley Davis, *Introduction to Total Quality: Quality, Productivity, Competitiveness* (Columbus: Macmillan College Publishing Company, 1994).

Gain the Commitment
of Top Management

═══ **MAJOR TOPICS** ══

■ Total Quality as Cultural Change
■ The Necessity for Commitment from the Top
■ Leadership by Example, Followership by Observation
■ Direct Involvement
■ Commitment of Time
■ Commitment of Resources

Of all the factors that contribute to successful implementation of Total Quality, the commitment of top management is first and foremost. Why this is so and what the commitment of top management entails are explained in this chapter.

TOTAL QUALITY AS CULTURAL CHANGE

Historically, Western organizations, whether in the public or private sector, have promoted internal competition and, in turn, adversarial relationships. Typically organizations have promoted, or at least condoned, competition—among departments, among employees, or between management and labor. In such organizations, a "me-against-you" attitude is part of the organizational culture. In many Western organizations this type of culture has led to an attitude among employees of "I can't win, so I'll just put in my time and collect my pay." In such organizations, employees and managers alike often complain that if only higher management would listen, work could be done more efficiently. At the same time, many upper-level managers in such organizations believe that unless employees are taken by the hand and told what to do, when to do it, and how to do it, they won't work at all. In these traditional organizations, few workers or mid-level managers have a firsthand knowledge of the organization's vision, goals, or objectives.

Consequently, they work in the dark, operating on the assumption that what they do every day is somehow the right thing to do. But this belief can be difficult to sustain when employees know from firsthand experience that there is a better way. Knowing this but not being able to do anything about it can't help but hurt morale. Total Quality offers a better approach.

Although Total Quality is not fully developed in the west, it is hardly a new concept. The term *total quality control* was first introduced by Dr. Armand V. Feigenbaum in his book, *Total Quality Control: Engineering and Management*, published in 1961. Feigenbaum, who was manager of quality control and manufacturing operations at General Electric, saw total quality control (TQC) as a system for integrating the quality development, maintenance, and improvement efforts of an organization so as to ensure production and service at the most economical levels that will guarantee full customer satisfaction. Feigenbaum recognized that TQC required the participation of many, although not all, departments. He believed that TQC had to be led by quality control specialists.

The Japanese had embarked on a similar course in 1950, principally under the leadership of Dr. W. Edwards Deming. There was a major difference between the Japanese approach—originally called Japanese-style total quality control—and Feigenbaum's approach. The Japanese never saw total quality control as the domain of quality control specialists, but instead trained all levels of all disciplines for participation in quality initiatives. In 1968 Japan began using the term *company-wide quality control* as the designation for their approach.[1]

Feigenbaum's approach, while certainly advancing the cause of quality and customer satisfaction, was not as comprehensive as the approach taken by the early Japanese advocates under the leadership of Deming. Convincing traditional managers to accept Feigenbaum's limited approach is difficult enough, but winning them over to Deming's much broader philosophy—one that requires a whole new organizational culture—is a real challenge.

Kaoru Ishikawa believed that organizations transform themselves by adopting six new ideas.[2]

1. Quality, not short-term profit, first.
2. Consumer, not producer orientation. Think from the standpoint of the customer.
3. Elimination of internal barriers to quality (departmentalism).
4. Utilization of statistical methods.
5. Respect for people as a management philosophy; full participation.
6. Cross-functional management.

THE NECESSITY FOR COMMITMENT FROM TOP MANAGERS

The Total Quality philosophy requires that all elements of an organization be involved actively and constantly in the new way of doing things. It is not sufficient to implement Total Quality in a single department of a company; by definition, such an approach would not be Total Quality. Imagine an organization that has six departments headed by six different department managers, all of whom are of equal rank and report to the same

---

TOTAL QUALITY TIP

The Notion of Culture

"The notion of culture in a company is complex and elusive. In general, culture refers to the everyday work experiences of the mass of employees. We urge managers to address these questions: How do employees experience their jobs? What gets in the way of employees' pride in their work and their work groups? Do they feel valued and trusted by the company?"[3]

Peter R. Scholtes

---

general manager. Who has the authority to implement Total Quality throughout the entire business? Certainly none of the department managers. Only the general manager to whom they report has the authority. If this organization is traditional in its approach to doing business, each of the departments has its own agenda, one which may be somewhat in step with that of the overall organization, but which is primarily designed to promote the department's own growth and thereby ensure its own welfare.

Such departmentalism can work to the detriment of the overall organization by misdirecting energy and resources. They represent a serious impediment to the successful implementation of Total Quality. Too many general managers believe that a little healthy competition among the departments in their organization is good for business. This is an unfortunate misconception. Imagine a football team in which the offensive line has a different agenda than that of the backfield. This team's offense would not be very productive. Football teams train constantly to ensure that all of their subunits have the same agenda—winning the game—and that they all work together in pursuit of that agenda.

If ever there was a need for total cooperation, it is among the departments within an organization. Competitive energy and resources should be directed against an organization's competitors. Figure 1–1 illustrates the effects of internal cooperation versus internal competition. In the figure, Organization B is less likely than Organization A to achieve the desired result. Nevertheless, this is the way many organizations operate, and department managers lack the power to change it. Only the top manager can do that. Only the person at the top of the organization has the power to see to it that all departments become actively involved in Total Quality in a positive manner. This is why commitment from the top is not just important, it is essential.

LEADERSHIP BY EXAMPLE, FOLLOWERSHIP BY OBSERVATION

Leaders are supposed to inspire their people to superior performance, channeling them, in spite of their disparate personal objectives and motivational factors, into a common effort that is greater than the sum of their individual capabilities. The question is, how? There are many ways, as illustrated in Figure 1–2.

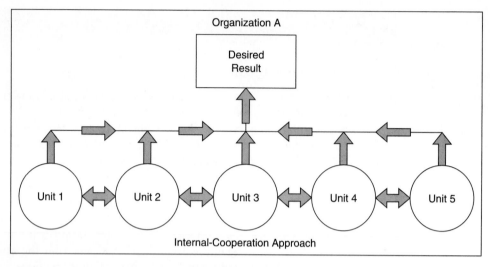

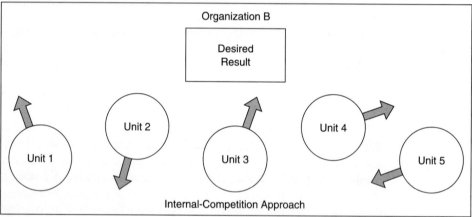

Figure 1–1
Internal Cooperation versus Internal Competition

The rousing Churchillian call to the challenge—the motivational/inspirational approach—is one way. Another way is the low-key approach, which relies on persuasion. Regardless of the approach used, *trust* is an essential ingredient. Followers must believe what leaders tell them and be confident that they will be supported if the venture goes sour. The most powerful approach to leadership is *leadership by example, followership by observation*. With this approach, people observe the leader's actions and emulate them or work in ways to support them. This is the crux of leadership and commitment from the top.

Figure 1–2
How Quality Leaders Lead

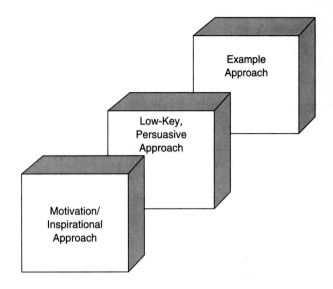

ACHIEVING COMPLETE MANAGEMENT COMMITMENT

As illustrated in Figure 1–3, complete management commitment involves direct involvement, commitment of time, and commitment of resources.

Direct Involvement

Not only must senior managers state their commitment, they must demonstrate it by actively participating in the Total Quality effort. Only when employees see their leaders involved in Total Quality activities will they accept that a commitment does, in fact, exist. People follow not just by hearing, but by observing. This point is lost on many managers, but it is critical for the implementation of Total Quality.

Commitment of Time

Ask executive-level managers to name their most valuable resource and many will respond with one word: time. Few people have enough time to accomplish everything

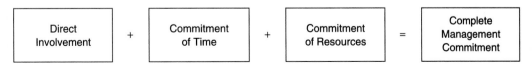

Figure 1–3
Elements Contributing to Complete Management Commitment

TOTAL QUALITY TIP

The Participative Management Process

"The Participative Management Process (PMP) provides Motorola with a perspective that is quite different from mainstream corporate America. The PMP rests on four cardinal principles. . . . management commitment, participation, communication, and trust. The most critical of these principles is management commitment, because it is central to the success of the other three principles."[4]

James A. Belohlav

---

they want to do. Often higher-level managers find their limited time consumed by meetings. Typically these are very busy people with other things to do but too little time to do them. Effective delegation can help, but there are still many responsibilities competing for the executive's attention.

When Total Quality is proposed, executives are asked to commit even more of their already limited time. Further, they are told that the responsibility for Total Quality cannot be delegated. Many proposed total-quality initiatives never get past the discussion phase because top executives and their staffs do not feel they can commit as much as half of their time to the effort. This reaction is both unfortunate and premature, because as soon as Total Quality begins to take hold, fewer demands are made on executives. This is because decisions get made farther down in the organization and many problems are prevented that used to require the executives' attention. Nevertheless, executives who already work long hours are often reluctant to make a commitment that promises to take even more of their time. How can this hurdle be overcome? In many cases it is not overcome. Top executives often decide that neither they nor their staffs have the time to implement Total Quality. In other cases they may conclude that Total Quality is something they really want, but that it will have to wait until they have more time to devote to it. In both cases, the result is a bad decision. The fact that executives are working such unreasonably long hours is evidence that the organization is not working well, and that Total Quality is needed badly.

The belief that things are somehow going to get better in a few weeks or months is unrealistic. The only real way to overcome the time hurdle is by making an honest, objective appraisal of the situation. How many of the things that consume executives' time are really important to the business? How many things could be handled as well as or better by subordinates? How many things are not worthy of anyone's time? By asking themselves the questions listed in Figure 1–4, executives can identify activities that are not meaningful and can be eliminated, and others that can be delegated to a subordinate. Typically, such an approach frees up the time executives need to implement Total Quality.

Executives who have successfully implemented Total Quality relate that initially they spent a third to half of their time involved with total-quality activities. These same exec-

- Are you making decisions that could be made by people who are closer to the problem?
- Are you reading reports daily that could be read weekly? Are you reading reports weekly that could be read monthly?
- How much of your time is spent in meetings that take too long to accomplish too little?
- Have you considered having all meeting participants stand up? This would result in shorter meetings that still accomplish their purpose.
- Are you making wise use of the time you spend driving or commuting (i.e., could you be doing dictation, returning telephone calls, reading reports, etc.)?
- Is your paperwork routinely organized for maximum efficiency?
- Are you making effective use of time-saving technologies (i.e., personal computers, special telephone features, FAX machines, microcassette recorders, etc.)?
- Are you cutting down on travel by using alternatives such as video conferencing where appropriate?
- How often do you review your daily activities and ask yourself how each activity contributes to accomplishing the organization's goals? Are you wasting time on nonessential activities?
- Do you restrict staff memos to one page and require the *bottom line* to be stated in the first sentence?

Figure 1–4
Time-Saving Questions for Executive Managers

utives eventually spend 100 percent of their time involved in total-quality activities because the Total Quality approach is how they now do business. In other words, everyone in the organization is doing things the Total Quality way. The executives relate that they now have time to do the things that need to be done, such as planning for the future, talking and listening to employees, responding to ideas and concerns, and intelligently leading the business.

Total Quality requires a second level of commitment from executives that is also critical. Not only must top executives commit their own time, but the time of all employees in the organization. Employees will be involved in training, team activities, preparing reports, making recommendations for continuous improvement, and all other total-quality activities. Why must this commitment of time come from the top? Because if it doesn't, it is unlikely that middle managers will give their employees the time to get involved. When there is a commitment from the top, although there may still be some grumbling among the middle managers, employees will be allowed to participate.

The important concept to grasp here is that although there is an initial time requirement on the part of the high-level managers, this time can be considered an investment in the future, both for the long-term benefit of the organization and the reasonably near-term benefit of the managers who invest the time. A successful implementation of Total Quality will result in fewer crises to be dealt with, fewer problems requiring top management involvement, and a smoother-running organization with fewer day-to-day demands on management at all levels. The ultimate payoff is that managers will have time to do real management work, such as leading, planning, facilitating, training, and developing their organization.

Commitment of Resources

It should be clear now that top management must make a complete commitment to Total Quality, must be directly involved, and must commit the time required. In addition to time, it is also necessary to commit resources. An organization's resources include its people, money, facilities, and equipment.

An eventual result of Total Quality should be cost reduction. Although the implementation of Total Quality does not necessarily carry a high price tag, there will be costs associated with it. These costs typically include the following:

- Cost of providing training for the Steering Committee and all other employees
- Consulting and technical assistance
- Pay for employees who are engaged in Total Quality activities

Top managers serving on the Steering Committee will require training. Occasionally top managers conclude that Total Quality is simply a catchphrase for good management, and therefore that they need no training. A reasonable response to this interpretation is, "If Total Quality is just good management, why haven't you been managing this way all along?" Managers tend to manage either as they have been trained to manage, or as they themselves have been managed. Consequently, traditional managers need training in Total Quality whether they think so or not. The training should include not only the philosophical and conceptual elements of Total Quality, but the practical ones too. For example, the quality tools and their applications to problem solving should be part of the training for even the higher-level managers. At the Steering Committee level, it is also essential to include team building in the training. Steering Committee members are usually high-powered people who have spent their careers achieving as individuals, and they will need to learn how to achieve as team members.

Training will also be required for all employees before they begin their involvement in Total Quality. This training should include team building, problem solving, and using quality tools. Training and other implementation activities will take employees away from their normal duties for an hour or more a week. Since the employees continue to be paid while participating in these activities, the costs associated with their time must be considered.

The organization should have a facilitator on each team to assist team members as they develop. Facilitators also will have to be trained and paid for the time they spend on their facilitation duties.

Most organizations will need a consultant to help with the initial phases of implementation and to coach as needed throughout the process. While it is possible to spend a great deal for consulting services, it is by no means necessary. Competent consultants who work for reasonable rates are available in most locations. A good consultant will save the organization far more than he or she costs.

The money to pay for all these activities must be available when it is needed—another reason that commitment from the top is essential. A difficulty that must be faced here is that with so many factors that affect a company's performance, it is nearly impossible to project the payback on the organization's investment in Total Quality. For example, it is difficult to know with certainty that the money invested in training resulted in specific performance gains. Deming was correct when he held that some things are simply not measurable. The test to apply before committing money is reasonableness. Does it make sense? Is the timing right? Is the money available? Can we afford it? Is it the right thing to do? If the answer to each of these questions is yes, the commitment makes sense. The commitment of resources is just as important as the commitment of time. Without both, the implementation of Total Quality will fail.

MAINTAINING MANAGEMENT COMMITMENT TO TOTAL QUALITY

Managers should be wary of the tendencies to lose their initial enthusiasm over time and to lapse into old, familiar management patterns. The list of questions in Figure 1–5 can be a useful device for managers and leaders to periodically test their continued commitment to implementing the Total Quality approach.

Results won't be immediate, but after a few months of sustained, concerted effort, there will be solid evidence that Total Quality is causing significant improvements throughout the organization, improvements that will result in better quality, greater satisfaction among customers (both external and internal), improved business performance, and a more competitive enterprise.

SUMMARY

1. Implementing Total Quality requires cultural change in an organization. This change can be summarized as follows: think quality, not short-term profits; think from the standpoint of the customer; break down internal barriers to quality; use statistical methods to measure performance; and involve employees in decision making.
2. Implementing Total Quality requires complete commitment from the top. The resources needed and the internal cooperation required will come only when top management makes a total commitment and gets personally involved.
3. It is critical that executive-level managers model the total-quality behaviors they want employees to emulate. This is called leadership by example, followership by observation.

- Do you set an example of commitment to total quality or do you just talk about it?
- Do you find ways to make total quality work or do you make excuses for letting it fail?
- Are you open to trying new ideas, approaches, and strategies or do you automatically react with cynicism?
- Do you take personal responsibility for seeing that improvement efforts succeed or do you assume someone else will do what has to be done?
- Do you deliver more than you promise or do you promise more than you can deliver?
- Do you continually challenge how things are or do you take comfort in doing things the way they have always been done?
- Do you involve and empower others or do you take the "lone hero" approach?
- Do you create conditions that allow change to happen or do you expect people to change in spite of existing conditions?
- Do you enable people by providing the resources needed for success or do you expect people to succeed without the necessary resources?
- Do you inspire people through personal, face-to-face interaction or do you remain isolated and aloof?

Figure 1–5
Commitment Test for Managers and Leaders

4. One of the major commitments executive management must make when implementing Total Quality is the commitment of time. This means a commitment of their own time as well as that of all employees involved in the implementation.
5. Another major commitment executive management must make is the commitment of financial resources. There are costs associated with implementing Total Quality. They include the cost of providing training for the Steering Committee and all other employees, consulting fees, and pay for employees who are engaged in implementation activities.

KEY TERMS AND CONCEPTS

Commitment from top managers	Leadership by example, followership by observation
Commitment of resources	
Commitment of time	Low-key, persuasive approach to leadership
Cultural change	Motivational/inspirational approach to leadership
Departmentalism	
Full participation	Steering Committee

REVIEW QUESTIONS

1. Explain what is meant by cultural change as it relates to implementing Total Quality.
2. Why is commitment from the top so important when implementing Total Quality? Give an example that illustrates this point.
3. Explain the concept *leadership by example, followership by observation* as it relates to the implementation of Total Quality.
4. How does commitment of time figure into a Total Quality implementation?
5. How does commitment of resources figure into a Total Quality implementation?

ENDNOTES

1. Kaoru Ishikawa, *What Is Total Quality Control? The Japanese Way.* (Englewood Cliffs, NJ: Prentice-Hall, Inc., 1985), 90.
2. Ishikawa, 104.
3. Peter R. Scholtes, *The Team Handbook* (Madison, WI: Joiner Associates, Inc., 1992), 1–16.
4. James A. Belohlav, *Championship Management: An Action Model for High Performance* (Cambridge, MA: Productivity Press, 1990), 64.

IMPLEMENTATION CASE STUDIES

Each chapter in this book concludes with two case studies that relate directly to the phase of the implementation covered in the corresponding chapter. One case describes an organization that takes the right approach to implementing Total Quality. The other case illustrates the wrong approach to implementation. In each chapter, the Manufacturing Technology Corporation (MTC) takes the right course of action, while the Engineering Services Corporation (ESC) takes the wrong course of action.

MTC employs 400 people and manufactures electronic products for military applications. ESC employs 300 people and provides engineering services in the following disciplines: mechanical, electrical, civil/structural, chemical, environmental, and aeronautical engineering.

CASE STUDY 1–1

Commitment from the Top at MTC

Never in its history had MTC needed so badly to improve its competitiveness. The drastic cuts in military contracts that followed the fall of the Soviet Union had eaten deeply into one of MTC's largest markets. Clearly, two things had to happen. First, MTC needed to hold onto the largest possible piece of the rapidly shrinking military pie. Second, MTC needed to expand into non-military, or at least dual-use, product lines. Even a cursory assessment of MTC's capabilities revealed that the company faced an uphill battle.

MTC's personnel were geared toward being the low bidder on government contracts. Change orders and cost overruns were an accepted part of doing business. Quality, cost, productivity, and response time had been only tangential issues. Now, overnight, or so it seemed, these issues had become not just critical, but key to MTC's survival.

CEO John Lee knew his only chance was to take charge of the situation with bold action. He decided to place his hopes for survival on Total Quality and to make a 100 percent commitment to its successful implementation. Lee called a meeting of his top managers and made the following statement:

> "Ladies and gentlemen, MTC is headed for bankruptcy. On our current course, we will be out of business in two years or less. If we are going to save this company, our jobs, and the jobs of 400 employees, radical change will be necessary. I have asked you here today to say that we are going to make the necessary changes. We are going to implement Total Quality as the way we do business at MTC. I am making a total commitment to the implementation and asking you to do the same. You and I will be responsible for ensuring the success of the implementation. The future of this company is in our hands."

Lee dismissed the meeting without allowing any discussion. Instead, he gave each manager a copy of the book *Introduction to Total Quality,* asked them to read it, and to meet again in one week prepared to discuss what Total Quality and a 100 percent commitment to it will mean to MTC.

CASE STUDY 1–2

Lack of Commitment from the Top at ESC

Engineering Services Corporation (ESC) had been Amanda Bornstein's only child. She had started the company in her garage forty years ago when, as the only female engineering student in her graduating class, the best employers either passed over her completely or offered a starting salary much lower than she deserved. Several of these companies had paid a high price for their shortsightedness in lost business over the years. Bornstein was an outstanding engineer, a tough competitor, a shrewd judge of talent, and an excellent manager. When she sold ESC and retired, Amanda Bornstein could relax knowing that she had built a world-class company with an international reputation for quality. Now, just three years later, the company was in trouble. Business had fallen off by 30 percent and the list of unhappy customers grew almost daily.

In desperation, ESC's CEO, John Hartford, had asked Bornstein for help. She spent a week talking to employees, long-time customers, and suppliers before giving Hartford her analysis of the situation. She summarized her findings as follows:

- ESC had lost its focus on the customer.
- Adversarial relationships had developed with suppliers and among internal departments.
- Interaction among managers and employees had become intermittent at best.

- Employee morale had dropped to an unprecedented low and there was no longer a sense of being a member of a team.

Bornstein recommended a return to the Total Quality approach to doing business she had used to build ESC even before anyone had heard of Total Quality. Hartford had agreed that implementing Total Quality made sense, but Bornstein was worried. So far his actions had not matched his words.

Hartford called his top managers together and shared Bornstein's analysis of the situation with them. However, what was supposed to have been a discussion of the need for major changes quickly degenerated into a shouting match in which the participants pointed the finger of blame at each other. Hartford finally called the meeting back to order and made the following statement:

> "We are going to have to make some changes around here or we will all find ourselves in the unemployment line. That new kid in the Mechanical Engineering Department, Lane Watkins, is always talking Total Quality this and Total Quality that. I think he picked it up at the University of Florida. ESC is going to go with Total Quality, and I'm putting Watkins in charge of the implementation. I know he's just a kid, but he knows more than any of us about Total Quality. I want everyone in this room to give him your cooperation. Now let's get back to work."

When Amanda Bornstein reflected on how Hartford handled the meeting, she couldn't help but worry about the company that had been her life. Was Hartford really committed to Total Quality, or was he just going through the motions?

Create the Total Quality Steering Committee

The hierarchical organization that is typical today has been in use at least 3,400 years, and probably longer. The rationale for hierarchical organizations is simple. As any organization grows, it becomes too large for one person to manage. Therefore, responsibility must be divided and delegated to each division. When the divisions get too large for their managers to oversee, they too are split into departments, each with its own department head, and so on. In this way, the intermediate managers share the work load of the person at the top. This is the concept of delegation. At least this is the way it is supposed to work. Ordinary people among the tribes of Israel had to work their way up through three or four layers of bureaucracy before they could make a demand on Moses's time. Moses intended that their problems be solved as far down in the organization as possible (a very modern concept), so he could devote time and energy to the larger issues. The hierarchical organization was surely an improvement over the chaos that must have prevailed in the wilderness before, but it could only be as effective and efficient as the people who were put in the positions of responsibility.

Moses had his troubles with the hierarchical organization, and we have ours. Not only is there the issue of the competence of the people in the organizational structure, but just as often there is the problem of their having diverse agendas. If the people in an organization are not all working toward the same goal, the performance of the total organization suffers. Sometimes managers focus more on their unit's interests than on

---

TOTAL QUALITY TIP

The Traditional Hierarchical Organization

"You are a burden too heavy for me to carry unaided. . . . So I took men of wisdom and repute and set them in authority over you, some as commanders over units of a thousand, of a hundred, of fifty, or of ten. . . ."[1]

Moses, c. 1400 B.C.

---

those of the overall organization. Sometimes higher-level managers do not let unit managers see the big picture. When the desires of the leader are not known throughout the organization, middle managers are forced to guess about expectations. Following leaders can be difficult enough when subordinates know where they are heading, but it can be impossible without that knowledge. This chapter does not propose the elimination of a hierarchical organizational structure, but it does propose a different way of leading organizations. Organizational structure is addressed in a later chapter.

THE DIFFERENCES BETWEEN TEAMWORK AND STAFF WORK

In a traditional organization, unit-level managers are concerned almost exclusively with their own units. For example, the head of an engineering department sees his or her job as assembling the right people to do the engineering job, providing them with the tools and facilities they require, and leading them in the accomplishment of the engineering tasks assigned to their unit. This person's interest may stop at the walls of the engineering department. He or she is not likely to feel any responsibility for the manufacturing or purchasing departments. Such managers tend to have the attitude "You do your job and I'll do mine." Any department in a traditional organization could have been used for this example. Departments in this kind of organization tend to put walls around themselves and become self-contained. These walls can become impenetrable and, as a result, inhibit communication among departments.

Often organizational units are too large to be managed by one person. This is why departments are formed. As this happens, the department heads become members of the leader's staff. Staff members are the people with whom the leader works most closely. Frequent staff meetings are common in hierarchical organizations. These meetings allow leaders to communicate with all of their staff members at once, afford staff members the opportunity to communicate with the leader and the whole group, and facilitate decisions involving more than one staff member and the leader. The dynamics of discussion depend on the leader and his or her style. Some leaders actually squelch discussion by being critical or judgmental. Unfortunately, aside from periodic staff meetings, it is astounding how little interaction there is among department heads in many traditional organizations.

TOTAL QUALITY TIP

Organizations Are Made Up of Teams

"An organization is one large team with many sub-teams. Each team works together to keep the organization competitive and to accomplish its overall mission."[2]

Stephen Uselac

The way many organizations work is that department heads get their instructions and do their best to carry them out, but in an isolated manner without regard for what the other departments are doing or how they fit into the larger system. Such situations are examples of working in a vacuum—the worst possible approach for organizations trying to compete in the global marketplace.

It is possible to correct such situations without changing the organization at all. You simply change the roles of department heads so that they become team players with the other department heads and the leader. When it comes to implementing Total Quality, the most successful organizations are those in which the leader's staff serve as the Total Quality Steering Committee. The difference in how staffs and Steering Committees operate are subtle and may seem minor at first, but a closer examination will reveal major differences:

- In traditional organizations, decisions are typically compartmentalized and made by the leader and the appropriate staff member. In Total Quality organizations, the entire Steering Committee gets involved in all decisions, regardless of whose department they apply most directly to.

- In traditional organizations, instructions have to be tailored to the appropriate department. In Total Quality organizations, a single objective can be set for the departments to achieve together.

- In traditional organizations, it is difficult to involve more than one department in a project due to turf problems and parochial tendencies. In Total Quality organizations, all the department heads share responsibility, resulting in greater cooperation.

- In traditional organizations, it is often unclear how a project relates to the vision, goals, and objectives of the organization. In Total Quality organizations, the relationship becomes clear through Steering Committee dialogue and the corresponding communication within and among departments.

As members of the Steering Committee, staff members must function as a team under the leadership of the top executive (the Leader). In this way, they focus on what is good for the entire organization rather than what is best for their individual depart-

Figure 2–1
DATAMAX's Expenses for 1993

DATAMAX	PL-G ($000)	PL-A ($000)	Division Totals ($000)
Mfg. Direct Labor	500	500	1,000
Variable Overhead	200	200	400
Fixed Overhead	750	750	1,500
Subtotal	1,450	1,450	2,900
G&A*(@25%)	363	363	725
Cost of Goods Sold	1,813	1,813	3,625

* General and Administrative

ments. The following example illustrates how a department in an organization can win, but in so doing cause the overall organization to lose.

DATAMAX,* a division of EMP Corporation, supplies military hardware to the government. Its main product is communication equipment. The division is set up with two product lines: ground-based and airborne equipment. Each division is headed by a Vice President who reports directly to the General Manager. A single manufacturing department produces the hardware for both product lines. In fiscal year 1993, manufacturing accounted for $1,000,000 in direct labor costs, evenly divided between Product Line A (airborne) and Product Line G (ground). Manufacturing carried $1,900,000 in expenses in its overhead pool, of which $400,000 was associated with people (fringe benefits, vacations, holidays, etc.) and $1,500,000 was related to facilities and equipment. This resulted in an overhead rate of 190 percent. In 1993, both product lines were equal in volume and both performed equally well, as shown in Figure 2–1.

In 1994, Product Line A was taken over by a new Vice President, John Lucky. Lucky began his new job by conducting an analysis to determine whether cost reductions were possible. He reasoned that anything done to cut costs would yield higher profits for his product line. Lucky felt that the best way to serve DATAMAX was to produce results in his product line, and he intended to do just that.

After the analysis was completed, it occurred to Lucky that he could save money by contracting out the manufacturing work for his product line. DATAMAX's manufacturing department was efficient enough that small outside contractors had no direct labor advantage, but they did have a significant overhead advantage. This meant they could offer lower costs than DATAMAX's in-house manufacturing department.

In 1994, Lucky convinced the General Manager to let him go outside for his manufacturing. Predictably, the year-end results met his expectations. He had reduced costs for the same products delivered a year earlier by $300,000. The General Manager was pleased with Lucky's performance. The only fly in the ointment for the General Manager

---

*DATAMAX and all other names used in this example have been changed. However, the example is real.

was the other product line. Its costs had increased over the previous year by $849,000. This wiped out the corporate gain made by Lucky and reduced overall profits by nearly $550,000 from the previous year on the same level of product output. Product Line G had done nothing differently. The only change was that manufacturing's overhead rate had gone from 190 to 340 percent. The direct labor demand for manufacturing had dropped by 50 percent, resulting in a major layoff, but there was no corresponding way to quickly liquidate facilities and capital equipment which represented most of manufacturing's overhead.

In an effort to understand what had gone wrong, the General Manager examined the books for the two years in question. The numbers for 1994 are shown in Figure 2–2. Upon comparing the 1994 numbers with those from 1993, the General Manager learned some interesting facts, some of them obvious, others more subtle. First, it was clear that Lucky's improved performance was almost entirely the result of avoiding internal overhead on his manufactured goods. He also avoided some G & A expenses. The proof performance of Product Line G had gotten worse, although the rate was actually lower by four points. It soon became clear that both of these differences were the result of Lucky taking his half of the manufacturing load outside to subcontractors.

When 50 percent of manufacturing's direct labor was taken away, the division's total overhead expense went down by only 10.5%—the $200,000 variable overhead associated with laying off half the manufacturing workforce. The 89.5 percent of the original overhead expense that remained now had to be applied for accounting purposes entirely to the product line that stayed in-house, Product Line G.

The numbers told the General Manager something he didn't want to hear: what is good for an individual product line may not be good for the overall organization. By going outside to subcontractors, Lucky had saved his product line $300,000, but he had

Figure 2–2

DATAMAX's Expenses for 1994

DATAMAX 1994			
	PL-G ($000)	PL-A ($000)	Division Totals ($000)
Mfg. Direct Labor	500	—	500
Variable Overhead	200	—	200
Fixed Overhead	1,500	—	1,500
Subcontract Labor	—	1,250	1,250
Subtotal	2,200	1,250	3,450
G&A*(@21%)	462	263	725
Cost of Goods Sold	2,662	1,513	4,175

* General and Administrative

cost the division a net loss of $550,000. What was good for Lucky's product line had been a disaster for the organization.

Now that he understood what had happened, the General Manager prepared a briefing chart and called his staff together. The manager of Product Line G figured he was about to be called to account for his results. He could take little comfort from the knowledge that his employees were performing just as well as they had in 1993. The primary difference was the huge increase in overhead that had been assigned to his product line.

All gathered at the appointed time, and the General Manager displayed the chart he had prepared, shown in Figure 2–3. The first column showed the total organization's numbers for 1993 and the second column showed the numbers for 1994. The third column, labeled *Cost Deltas*, showed the dollar differences, with + and – signs indicating whether costs were higher or lower in 1994.

The General Manager went through the chart line by line, comparing the two years. He emphasized that identical quantities of the same products had been produced in both years.

Direct Labor   In the first year, direct labor was $1,000,000. But in the second year, it was reduced to $500,000 by Lucky's decision to take his manufacturing work outside the company.

Variable Overhead   Variable overhead was also cut in half (to $200,000) by Lucky's decision to go outside for manufacturing, and by the corresponding layoff of manufacturing personnel.

DATAMAX	1993 (All MFG Inside) ($000)	1994 (1/2 MFG Outside) ($000)	Cost Deltas ($000)
Mfg. Direct Labor	1,000	500	−500
Variable Overhead	400	200	−200
Fixed Overhead	1,500	1,500	—
Subcontract Labor	—	500	+500
Subcontract Overhead	—	750	+750
Subtotal Production Cost	2,900	3,450	+550
G&A	725	725	—
Cost of Goods Sold	3,625	4,175	+550

Figure 2–3
Comparison of DATAMAX Expenses for 1993 and 1994

Fixed Overhead   There was no change in fixed overhead; it was $1.5 million in both years. Any changes in the fixed overhead here would take a long time due to fixed-length leases, depreciation rates on equipment, and so on.

Subcontracted Labor and Overhead   The General Manager didn't have hard data for the subcontractor's rates, but the rates were easily calculated with reasonable accuracy. What he did have was a price of $1,250,000 from the subcontractor whose job it was to produce exactly the same number of the identical products that had been produced in-house the year before. The in-house cost for direct labor was $500,000, so he made the assumption that $500,000 of the $1,250,000 was for labor. That left $750,000 for the subcontractor's expenses and profit. If that were true, then his overhead, G & A, and profit combined were 150 percent of labor. He felt he was close to the real numbers. But, regardless of how it was split, the price paid to the subcontractor was $1.25 million. So the deltas here were $500,000 for subcontractor direct labor and $750,000 for subcontractor overhead, G & A, and profit. Both figures were given plus signs since there had been no subcontracting the year before.

Subtotal Production Cost   Here was the key issue. The General Manager pointed out that in each of the two years, exactly the same production was achieved. However, in the second year production had cost the division $550,000 more. Since the products were sold for the same price, the organization's profits had to absorb the difference. Everyone quickly grasped the significance of this. They would all feel the effect of this poor performance in their incentive bonuses, which were tied to the organization's profits.

General and Administrative Costs   The General Manager pointed out that there had been no change in G & A expenses from 1993 to 1994, even though the manufacturing department had laid off half its employees. In both years the division incurred $750,000 in G & A expenses. The G & A rate had actually decreased from 25% to 21%. Unfortunately, it went down for the wrong reason. Since costs had gone up, it took a smaller percentage of the higher costs to amortize the G & A expenses.

Cost of Goods Sold   The General Manager called everyone's attention to the fact that the same $550,000 delta seen two rows earlier was also present in the bottom line. In 1993, the division's cost of goods sold was $3,625,000 and in 1994 it was $4,175,000 for the same goods delivered to customers.

Having shared the numbers, The General Manager asked for opinions as to what had gone wrong. The Accounting Director said it was obvious that the overheads had not been managed. If production had been halved, then overheads should have been halved as well. "Why couldn't we use the same argument for G & A?" asked the General Manager. He added, "We can't break leases, we can't just give capital equipment away to get it off our books, and we can't stop lighting and heating buildings because there are fewer people in them."

The manager of Product Line G, Lucky's counterpart, added his input. "From the numbers, it looks like when John took his manufacturing outside it left the division out of balance. Before this, we had equilibrium between labor costs and expenses. With half the labor gone, the fixed costs were still present and there was no way to avoid paying them. Unfortunately, there was nowhere else to absorb the expenses but my product line. When John went outside, he saved $300,000 for his product line, but only by shifting costs to mine." "That's right," said the General Manager.

The Human Resources Director ventured the thought that the $550,000 loss wasn't even the worst of the problem. "Twenty-five experienced employees—half the manufacturing workforce—have been laid off. Now that we have learned that outside manufacturing is not a gold mine, we don't have the workforce required to bring next year's production back in-house. We'll have to recruit, train, invest, and gain experience all over again." "That's right," said the General Manager, "but if we don't bring it back inside, the division is doomed because there is no way we can beat the competition with the kind of prices we'd have to charge in order to make money."

Lucky didn't like where the discussion was heading. He said he understood what had happened, but thought that the problem wasn't that his product line's work had been taken outside, but that the division had attempted to maintain a manufacturing capability when *all* the work should have gone outside. "Get out of the leases, get rid of the people. Manufacturing has just gotten too expensive."

The General Manager cut him off. "And what does the division have left when it dumps its manufacturing capability?" In marketing we talk about *differentiators*. That is, what would draw a customer to DATAMAX rather than to a competitor? Having a manufacturing capability separates us—differentiates us—from competitors who cannot build products in their own plants. Let's face it, anybody with a few good engineers can duplicate the functions of our products if the customers are willing to let them use a lot of off-the-shelf hardware that gets interfaced with a few subcontracted printed circuit boards. Our customers want more than that. They want equipment that is designed with their specific mission in mind and that pushes the envelope in performance, quality, reliability, and maintainability. If they can get what they want from a 'rack-em and stack-em' house, and sometimes they do, so be it. To be competitive in that environment, we'd have to take the division apart and start over again. And when we were done, we'd look just like a hundred others out there, all scratching for the same work that any garage shop in the country could win.

"DATAMAX has been successful for a good many years because we are vertically integrated, with Marketing at the front end, Engineering in the middle, and Manufacturing at the back end, making the product. Your suggestion is that we get rid of our manufacturing capability and all the expenses associated with it. Some might argue that such a move makes sense, and in the short run, perhaps it does. But why should we give business and profits to someone else when we can keep both within the company? We'll use subcontracting for work that we can't handle, either because we don't know how or don't have the tooling, or because we have peak loads that we can't cope with internally.

"From now on we're going to do any manufacturing we know how to do in-house. We'll bring Product Line A back in and we'll hire people to replace the ones we let go. The direct labor will put the overhead back in order again, and we'll go on from there.

"In the future, when we consider making drastic changes in the way we do business, we're going to look at it from every angle and we're going to know what's under every rock. If we had done that a year ago, the division would have been $550,000 richer, and all of you guys would have been better off–including Lucky."

THE RESPONSIBILITY ISSUE

In any discussion about the Steering Committee, the issue of responsibility is likely to be raised, usually by top level managers who sometimes mistakenly see the Steering Committee as taking away some of their authority. The Steering Committee is sometimes viewed by detractors as management by committee. Remember, a camel is said to be a horse designed by a committee. This negative view of committees usually occurs as the result of one or more of the following factors:

- Committees are used because top managers are unable to make decisions.
- Committees are a convenient way for managers to spread the responsibility for unpopular decisions.
- Managers believe that participation on a committee results in automatic buy-in on the part of participants.
- Aggressive staff members use committees to gain access and power they would otherwise not enjoy.

These factors should be thoroughly analyzed and understood. It is a mystery how people who are incapable of, or are habitually reluctant to, make decisions can advance through the ranks of an organization and eventually find themselves in responsible management positions. But it happens. One can only assume that in their earlier roles this trait was either not present or was well hidden. Regardless of how it happens, there is no shortage, in either the private or public sector, of managers who find decision making burdensome. Such managers tend to either delay making decisions by requesting more and more data, or they abdicate their responsibility and let subordinates or committees make the decisions. Either way, this is *management by abdication*, not *management by committee*.

Managers who try to avoid the unpleasant consequences of unpopular decisions by delegating them to committees are not leaders. Leaders must be willing to accept responsibility for unpopular decisions. Trying to spread the blame is not management by committee, it's *management by cowardice*.

There is much to be said for managing in a way that achieves *buy-in*. This is part of the rationale for employee involvement, a basic principle of Total Quality. Why suggest caution here, then? The answer is that Total Quality stresses teamwork. The Total Quality Steering Committee is a team. The kind of committee spoken against here is not a team. If a top manager's executive committee is made up of ambitious, hard-driving individuals, who either don't know or don't care about teamwork, buy-in is very unlikely to occur.

First, most traditional committees are made up of individualists. This is because of the way Western culture has historically picked winners, and it has been winning individuals who get promoted. Usually the people on an executive committee are there precisely because they excelled as individuals, not as team members. In such cases, real buy-in is uncommon. Knowing this, why does the Total Quality Steering Committee work? What is the difference?

The difference is that before the new Steering Committee takes any action, its members are trained to work as a team rather than as individuals who each march to a different drummer. Is it always possible to convert individuals to team players? No. This is an important point, and one that is seldom mentioned in the Total Quality literature. Sometimes the teamwork training does not work. It takes only one or two members who are unwilling to function as team members to subvert the efforts of the Steering Committee. This does happen. Dealing with such individuals begins with one-on-one counseling by the top executive. The message must be clear that although it is hoped the individual can make the transition to becoming a team player, if not, he or she will have to go.

A willingness to conduct such counseling sessions is part of the top executive's commitment. Often a staff member will see the reality of the situation and make the transition to becoming a team player. On the other hand, some people, for one reason or the other, will not make the change. When this occurs, such people must be retired, terminated, or otherwise removed. This is necessary because such managers cause an organization to be less competitive than it could be. Unfortunately, in many cases, organizations cannot bring themselves to remove senior-level managers. In such cases, these managers continue to be roadblocks to the Total Quality process and continue to cause the organization to be less competitive than it could be.

Committees are sometimes promoted by staff members who see it as their route to more access, power, or authority. Faced with a weak CEO, an aggressive staff member can use an executive committee to become the de facto boss. Through force of personality, connections throughout the organization, and their positions in various approval loops, such staff members can sometimes dominate decision making in the committee.

In the final analysis, no matter what the form of organization or management style, top managers remain responsible for all decisions, whether theirs or someone else's. Even when the group is operating as a team, the top executive remains responsible. The head coach of a football team is the one who has to answer for a bad season. He is the one whose job is on the line. The job is too big for one person, so he has a staff of assistant coaches, each responsible for some component of the team. These assistants work together as a team under the guidance of the head coach. One does not see assistant coaches working toward their own agendas, contrary to the head coach's general instructions. There is no question that the objective is winning the game, and that their energy and expertise are focused on that objective. Even so, the head coach still has overall responsibility for the team's performance. In this way the head coach and his coaching staff are similar to the Total Quality Steering Committee. The top executive retains responsibility for performance, but the Steering Committee is there to help him accomplish that objective. The Total Quality Steering Committee never represents management by abdication or management by committee in the sense discussed earlier.

ROLE OF THE STEERING COMMITTEE

The Steering Committee is similar to the coaching staff of a football team. Top executives cannot possibly attend to all the details of a business, make all the decisions, and execute all the functions. They must have staff members to do these things. Traditionally, functional managers have operated their respective departments within the organization, interacting with other departments within the organization only as required to accomplish specific tasks. The Total Quality Steering Committee broadens the roles of these managers.

In a Total Quality setting, staff members are no longer interested only in their departments. Each member must take an active role in the management of the organization, not just one department. Each member must become an assistant coach, supporting the coach's agenda and helping the rest of the team carry it out.

One of the first roles of the Steering Committee is to help the top executive establish the organization's vision. The Steering Committee translates the vision into a set of operating principles and broad objectives that clearly communicate the purpose and mission of the organization. The Steering Committee must try to understand the organization, including its strengths, weaknesses, attitudes, and concerns. The Steering Committee determines which activities should be undertaken and transforms individual departments into teams to ensure that their activities are consistent with the vision. The Steering Committee appropriately rewards the teams that meet expectations and assists those that fall short. In these ways, the Steering Committee manages the organization.

It is important to remember that the Total Quality Steering Committee includes the top manager. It cannot function without his or her personal, sincere, and constant involvement and active participation. Once Total Quality begins to take root, traditional management techniques become increasingly superfluous, and it is the Steering Committee that manages the organization as a team. The top executive is still responsible for the organization's performance, but he or she now has an effective coaching staff.

COMPOSITION OF THE STEERING COMMITTEE

The Total Quality Steering Committee is made up of the organization's top manager and the managers who report directly to him or her. This applies to all but the smallest organization, in which case a Steering Committee must be tailored to the organization's size and makeup. It is sometimes appropriate to add persons with specialized talent to the Steering Committee. For example, an organizational chart may show just three people reporting directly to the top manager, such as the heads of Marketing, Human Resources, and Operations. A Steering Committee with just three members would not make sense, because important functional departments would not be represented. In this case it would be advisable to bring in the managers of the departments under Operations: Finance, Engineering, Manufacturing, and Quality Assurance, as shown in Figure 2–4. The important point is that the Steering Committee should include the managers of all major departments, regardless of how the actual lines of authority are drawn.

It is not unusual for Steering Committees to include a permanent facilitator as an ex-officio member, particularly in the early stages of development. The facilitator's role is to keep discussion flowing, keep members on track, bring special training to bear on problem solving, and generally use his or her expertise to ensure that the Steering Committee functions as a team. While the facilitator can be very important, he or she does not have a vote, and for that reason is not considered a member of the Steering Committee.

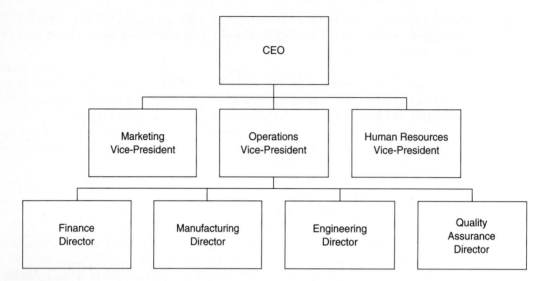

Figure 2–4
Example of Steering Committee Membership

THE STEERING COMMITTEE AS A WORKING TEAM

The role of the Steering Committee in managing the organization's various teams has been explained. This will be the Steering Committee's main focus early in its development. However, the Steering Committee is a team itself. Its members are trained in Total Quality and they are experienced, successful managers. A team made up of such people can perform a great service to the organization by using team methods to solve problems and continually improve processes.

Once the power of team problem solving and process improvement is recognized, it is natural for the Steering Committee to undertake such projects. Not only does the organization benefit from the collective expertise that gets focused on the issues, but equally important, a signal is sent through the organization that the top executives are applying the same methods the other teams are using and that they are involved. This lets employees know that the commitment to Total Quality is real, and not just a lot of talk about another half-hearted initiative that will soon run its course.

As the Steering Committee gains experience in functioning as a team, more and more of the staff functions and day-to-day management routines will be assumed by the Committee, leaving less to be done by individual members and department managers. A major result of this new approach is that the total organization will begin to operate as a single entity rather than a collection of semi-autonomous departments that seldom pull in the same direction.

Before the advent of Total Quality, most organizations set down some guidelines for their departments and left the day-to-day management of the departments up to the

---

TOTAL QUALITY TIP

Continuous Improvement and Teamwork

"Continuous improvement begins with team development and it can be sustained only by highly developed work teams. Improvement projects in the areas of customer satisfaction, work processes, and supplier performance can be designed and implemented successfully only by work groups that are able to function as teams."[3]

Dennis C. Kinlaw

---

respective department heads. As long as the departments more or less met their objectives (which may or may not have contributed to the achievement of the total organization's goals), the departments were left to their own devices. Whenever a department wanted to tap into the organization's limited resources—say, to purchase a new piece of capital equipment—competition for the available money would develop with other departments. Should Engineering request a new computer-aided design system, Manufacturing might counter that the money should be used to buy a new machine tool. Eventually the CEO would be asked to decide which expenditure best meets the organization's needs. In such cases there is always a winner and a loser. Interorganizational skirmishes such as this inevitably lead to morale-robbing animosities among functional managers.

Organizations with Total Quality Steering Committees that function as teams can avoid this scenario. The primary concern of the Steering Committee is the welfare of the whole organization. No advantage is gained by one department prevailing over another. Will there ever be disagreements among Steering Committee members? Of course. But as team members they will be better informed about the company's vision and objectives; they will understand the needs of all functional departments, not just their own; and they will be in a position that rewards the subordination of self-interest. Even if the ultimate decision is not the one hoped for, having complete information and being part of the process makes acceptance of the decision easier. No member is cast in the role of loser—or winner.

THE STEERING COMMITTEE AS A SYMBOL

Over the years there have been numerous quality-improvement initiatives. Many of them were devoid of substance, relying more on slogans than on commitment. In the early 1970s there was the zero-defects initiative. Ten years later the United States went through its quality-circles phase. Managers were told that quality circles were responsible for the enormous improvements Japan was making in quality and productivity. Companies rushed into quality circles without really understanding what they are or how to use them. Small wonder, then, that quality circles quickly failed in the United States. Having seen these and other programs come and go without living up to their promises, many managers and employees have become cynical.

The difference between these programs and Total Quality is substance. Total Quality is not program. It is a new way of doing business that is supported both by science and the psychology of human nature. In addition, there are many organizations worldwide that are living proof of its effectiveness, and the number is growing rapidly. Even so, skeptical employees must still be convinced that Total Quality is real. Few executives did anything with the zero-defects concept but talk about it. Even fewer executives got directly involved in quality circles. A hands-off approach by executive management is a prescription for failure. This is why executive involvement that is personal, sincere, and continuous is an absolute necessity when implementing Total Quality. When employees see the General Manager and Steering Committee members functioning as a team, behaving as if they believe in the process, getting results, using the same problem-solving tools taught to other employees, and acting as if they have the success of the total organization in mind, employees get a strong message that Total Quality is not another here-today, gone-tomorrow program. They see that management is serious and committed. This symbolic affirmation of Total Quality can do more to ignite the enthusiasm of the rest of the organization than all the speeches one could make.

SUMMARY

1. The traditional staff is a group of individuals who behave and interact as individuals, each with his or her own agenda. A Steering Committee functions as a team. Its members work as a single unit toward the same goal. Individual members are mutually supportive and all work on the same agenda.
2. Establishing a Total Quality Steering Committee focuses more energy and more brain power on improving the performance of the organization. However, it does not relieve top managers of their responsibility for the organization's performance.
3. The role of the Steering Committee is similar to that of a coaching staff. It sets an example of personal commitment, sets priorities, provides guidance, and establishes, monitors, and assists teams.
4. The Steering Committee is composed of the top manager and the heads of all major functional departments in an organization. The CEO's staff can serve as the Steering Committee if all major department heads are included on the staff. If not, the staff should be augmented with the appropriate personnel.
5. The Steering Committee works as a team. It can apply these same teamwork principles in helping continually improve processes throughout the organization. The Steering Committee is also a symbol. It will be watched by all employees to determine if top managers are really committed to Total Quality.

KEY TERMS AND CONCEPTS

Management by abdication	Responsibility
Management by committee	Steering Committee
Management by cowardice	Teamwork
Permanent facilitator	Top manager

REVIEW QUESTIONS

1. Explain the difference between teamwork and the traditional approach to staff committee work.
2. Confirm or refute the following statement: "A committee is just a weak manager's way of avoiding responsibility for difficult decisions."
3. What is the role of the Total Quality Steering Committee?
4. Who should serve on the Steering Committee?

ENDNOTES

1. *The New English Bible,* Deuteronomy 1:9–15.
2. Stephen Uselac, *Zen Leadership: The Human Side of Total Quality Team Management* (Londonville, OH: Mohican Publishing Company, 1993), 19.
3. Dennis C. Kinlaw, *Continuous Improvement and Measurement for Total Quality: A Team-Based Approach* (San Diego, CA: Pfeiffer & Company/Business One Irwin, 1992), 75.

CASE STUDY 2–1

A Well-Steered Steering Committee at MTC

After waiting a week, John Lee, CEO of MTC, called a staff meeting to take the next step toward implementing Total Quality. His staff members had read the book, *Introduction to Total Quality*, that Lee had provided them. Lee began the meeting by asking each staff member to comment on the ramifications for his or her department of making the transition to Total Quality.

The ensuing discussion was lively and at some points heated. One staff member expressed concern about the time requirements. Another revealed that she had doubts about the ability of employees to learn how to use basic quality tools such as Pareto charts, fishbone diagrams, and control charts. Others liked the idea and thought they should get started right away. However, it was the head of Accounting who best summarized the consensus of the staff when he said, "John, I don't see that we have any choice. We either change the way we do business around here or I've got to begin getting our books in order to begin a bankruptcy proceeding. I don't know whether or not implementing Total Quality will save us, but I do know this: success or failure will be determined by the people in this room. . . . "

John Lee moved the discussion from the "Should we do it?" to the "How do we do it?" stage by asking for input concerning the composition of the Steering Committee. The CEO's staff currently comprised the directors of the following departments: human resources, engineering, marketing, manufacturing, accounting, purchasing, and quality. Since each functional department is represented on the CEO's staff by its manager, the consensus of the group was that the existing staff should be MTC's Total Quality Steering Committee.

The Director of Purchasing asked about the use of a permanent facilitator. Lively discussion ensued as to the pros and cons of this approach, but ultimately there was little interest in using a facilitator. Several allusions to baseball and football were made during the discussion. The Director of Engineering, who had been a fast-pitch softball star in college, commented that she saw the Steering Committee as being MTC's coaching staff and that John Lee had to play the role of head coach. This was the consensus of the group.

Lee closed the meeting with the following statement: "As of today, we are MTC's Total Quality Steering Committee. The success of Total Quality at MTC is on our shoulders as is the survival of this company. Keep your copy of *Introduction to Total Quality*. I am contracting with its authors to train us as the Steering Committee and to coach us through the implementation."

======= **CASE STUDY 2–2** =======

Who Is Steering the Steering Committee at ESC?

John Hartford was beginning to question his choice of Lane Watkins as his in-house Total-Quality guru. He had met with Watkins and told him to get started with the implementation, but Watkins had balked. He didn't seem to understand what Hartford expected of him. Watkins saw himself as a facilitator who would help the Steering Committee, composed of Hartford and his top managers, implement Total Quality.

Hartford didn't want his top managers spending valuable time away from their jobs serving on a Steering Committee, and he certainly wasn't going to tie up more of his own time with another responsibility. He barely had a spare moment in his day. How could he justify spending additional time in meetings?

Watkins had pressed Hartford to establish a Steering Committee comprising his current staff augmented by three other functional managers. Watkins had stressed the importance of both commitment and personal involvement on the part of executive management. Hartford had listened patiently, but he just didn't agree. Finally, he gave Watkins his marching orders: "I am going to give you one representative from each department to serve on the Steering Committee. You will be the facilitator. Get them together and let them elect their own chairman. Don't worry about commitment. I'll tell the supervisors to cooperate with you and I'll come to your first meeting to show the Steering Committee that I'm committed."

Watkins didn't like this assignment, but he knew better than to press his luck with the CEO. After all, he was the new kid on the block. He would just have to do the best he could under the circumstances.

Build the Steering Committee Team

The third step in the implementation process is teambuilding for the Steering Committee. The Steering Committee will soon begin the formidable task of changing the culture of the organization. This is best accomplished as a team rather than a collection of individual department heads. Teambuilding activities typically require one to three days.

TEAMBUILDING DEFINED

To understand the concept of teambuilding, consider the example of training camp for a professional football team. While physical preparation is an important aspect of the training camp, mental preparation is equally important. Coaches use training camp to identify problems that might prevent the players from functioning as a cohesive unit or team. When problems are identified, they are dealt with immediately so that individual players can work together as a team before the season starts.

Some of the problems encountered in teambuilding are interpersonal. One team member may dislike or distrust another member. One team member may have difficulty communicating with or relating to another member. It is not necessary that all team members like each other, but it is important that they treat each other with trust and mutual respect. The team cannot function to its full potential unless interpersonal difficulties between members are eliminated. Trust, understanding, and a mutually supportive attitude are essential.

Another set of problems that may be encountered in teambuilding is associated with the play book (procedures) and game plan (strategy). Both have to be adapted to the skills and abilities of players or team members. For example, when developing a game plan for a football team, the passing ability of the quarterback, the speed and catching abilities of the receivers, the mobility and strength of the offensive and defensive lines, and many other factors must be considered.

Figure 3–1
Building Blocks in
Teambuilding

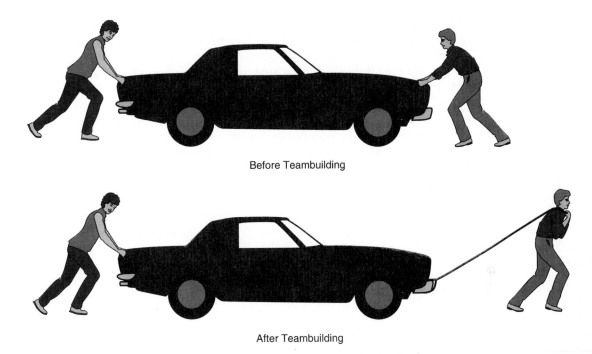

Before Teambuilding

After Teambuilding

Figure 3–2
Teambuilding Leads to Synergy

This is essentially what is meant by teambuilding. It amounts to identifying and solving interpersonal, organizational, and procedural problems within the team, thereby making it possible for individual members to work cohesively and effectively as a team. Figure 3–1 illustrates the elements involved in building an effective team.

THE GOAL OF TEAMBUILDING

In a Total Quality environment, work is accomplished primarily by teams of people rather than by individuals. The goal of teambuilding is to enable team members to work together toward a common objective, complementing each other with their skills, experience, and offsetting strengths and weaknesses. Teambuilding should lead to synergy, as illustrated by Figure 3–2.

WHEN IS TEAMBUILDING REQUIRED?

Teambuilding can help all organizations, even those whose staffs appear to work well together already. It would be a rare team that could not be improved. Unfortunately, many staffs do not work well together, and certainly many do not function as a team.

Western culture rewards individual achievement. Executives typically gain their high-level positions as the result of outstanding individual achievement, not because they are good team players. It is interesting to note that in the United States, the term "team player" typically connotes one who is willing to subordinate his or her family life, follow instructions, and never question the company's or the boss's wisdom or ethics.

One of the most difficult aspects of the implementation of Total Quality is converting a staff of individuals into a Steering Committee team. But this conversion is essential. Teambuilding should be a fundamental part of the implementation process in any organization. A checklist for determining whether teambuilding is needed is given in Figure 3–3.

BENEFITS OF TEAMBUILDING

If members of the Steering Committee function as kings of their own individual fiefdoms rather than a team, resources will be wasted on internal competition. In spite of the commonly held view to the contrary, internal competition is detrimental to the overall organization. By achieving teamwork at the Steering Committee level, the organization can eliminate internal competition and focus its resources on achieving the organization's goals.

Once Steering Committee members begin to function as a team, its members will find that their work has become easier and they have more time available. When the organization benefits, it follows that the Steering Committee members, along with all other employees, will also benefit. Executives who have created a Steering Committee that works as a team have said that they would never go back to the traditional way of managing and behaving.

If any one of the following situations is present, teambuilding is needed:

- High turnover in higher management
- Excessive work hours for higher management
- Quality problems
- Excessive rework/waste
- Frequent conflict
- Interdepartmental barriers
- Poor communication throughout the organization
- Employee attitude/morale problems
- Customer dissatisfaction
- Department agendas that supercede the organization's goals

Figure 3–3
Organizational Analysis Checklist

Figure 3–4
Environmental Characteristics
Needed for Teamwork to
Succeed

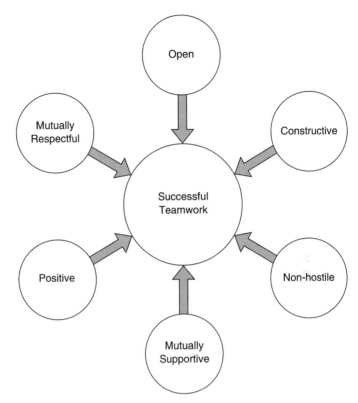

ENVIRONMENTAL CONSIDERATIONS

Figure 3–4 illustrates the environmental characteristics needed for teamwork to succeed. An open, constructive, non-hostile environment must be maintained throughout the process. Team members must be free and willing to state their opinions and observations honestly, openly, and completely. Holding back or using fuzzy language to disguise the real meaning of what is being said will prevent teambuilding from taking place. To be constructive, comments by team members should be restricted to issues related to working with the group or with individuals in the group. Comments on the personal characteristics of team members have no place in a teambuilding setting.

Team members should be mutually supportive. The only acceptable comments and observations are those intended to help others become more effective as team members. Comments should not be directed at tearing down or discrediting a team member. Mutual respect, honest concern, and awareness of personal sensitivities among team members are essential elements of teambuilding. The teambuilding facilitator must ensure that the discussion is not allowed to become personal or accusatory. If this happens, the facilitator should quickly guide participants back into a positive frame of reference.

In some cases the teambuilding facilitator is an outsider who has the skills to establish and maintain a positive environment, and who can act independently of the various individuals involved. In other cases the senior executive acts as the facilitator. When this approach is used, there is always the possibility that an attitude of "We'd better go along with the boss" will develop. If this happens, the team will be worse off than before. Senior executives who are ostensively serving as facilitators but who instead dominate the discussion and discourage participation inhibit real teambuilding.

It is critical that all participants believe that their input is desired; that they can be frank, honest, and open without risk to themselves; and that everyone can participate, with no individual dominating. The team's leader must be able to participate actively, openly, and in a nondefensive manner. The facilitator should maintain the interchange of thoughts, ideas, and comments at a positive, constructive, supportive level.

RESPONSIBILITIES OF PARTICIPANTS

The main reason it is difficult for the Steering Committee leader (the CEO) to also serve as the teambuilding facilitator is that the responsibilities of the two roles are so different. The leader is responsible for:

- Convincing the participants that he or she (the leader) honestly wants the ideas, thoughts, comments, suggestions, and solutions of the group.
- Refraining from a defensive or argumentative posture, even when comments are made about his or her behavior.
- Committing the necessary resources to teambuilding, including his or her time and that of the staff, facilities, and outside expertise, if required.
- Addressing the issues that are presented, even to the extent of modifying his or her own behavior when necessary.
- Supporting the long-term effort of teambuilding. The teambuilding session is just the beginning of long-term team development.

All members of the Steering Committee are responsible for:

- Behaving in a professional, nonaccusatory, nondefensive manner.
- Being honest and open in comments and making them clearly, without obfuscation.
- Participating fully.
- Having a desire to work together.
- Being committed to building up the team, not tearing down individuals.
- Keeping comments made in the teambuilding session within the group.
- Listening to the other members, thinking, and promoting understanding.

The Steering Committee's facilitator is responsible for:

- Ensuring participation by all members while preventing domination by any member, including the leader.

- Keeping the flow of comments and ideas positive and constructive.
- Heading off and preventing negative and defensive dialogue.
- Stimulating and guiding discussion through thoughtful questions and suggestions, and being careful to stay out of the way once the discussion is moving.
- Ensuring that problems, solutions, plans, and other significant items are recorded.

THE TEAMBUILDING PROCESS

One of the primary goals of teambuilding is to identify and solve the problems that prevent people from working together effectively. Such problems are usually found in two important areas: interpersonal relationships and organizational roadblocks, particularly work procedures. Although a particular staff supposedly works together in the normal pursuit of any enterprise, in reality there may be more work than togetherness. The issues that prevent them from working together in a harmonious, mutually supportive way may never have come up formally, but at the individual level, each person has his or her ideas of how the group's performance could be improved. An objective of the teambuilding activity is to get those issues out into the open, where they can be examined and resolved. As illustrated in Figure 3–5, this is done by exploring the Four W's:

Who?   Since the goal is to convert the Steering Committee into a team, it is important that all members be present throughout the teambuilding session. Teambuilding activities should not be attempted without the attendance of all members of the group. There

Figure 3–5
The Four W's for Identifying
Issues that Need To Be
Resolved

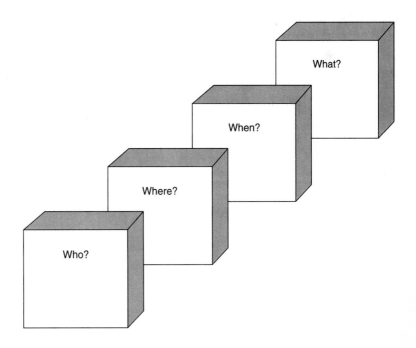

has to be complete buy-in and trust. This cannot happen for the individual who, for whatever reason, misses a session.

Where?   Teambuilding activities should take place in a relaxed, comfortable atmosphere, free of interruptions and daily comings and goings. The meeting place should provide sufficient privacy that participants will not be concerned about being overheard outside the room. It is customary to hold teambuilding meetings off-site, usually at a hotel or conference center. Good results can also be achieved when meetings are held in the home of a participant. The key is that for the duration of the training, whether a day or a week, the participants should be free of interruptions from their offices and should feel comfortable and secure in their surroundings. It also helps if meals can be served at the meeting site.

When?   The calendar date is not particularly important, except to the extent that the organization may be busier at one time of the month than at another. Select training dates so as to minimize conflict, and do it far enough in advance that conflicts can be worked out. The time of year can be a consideration if outdoor activities are planned, but don't delay teambuilding training too long waiting for good weather.

What?   Although a general agenda is desirable, it can be difficult to predict how long training sessions will take. The length of a session depends on the number of issues brought up and how difficult they are to resolve. The best approach is to work from a loose agenda that covers the principal elements of teambuilding without putting tight time constraints on the interactive parts of the session.

A typical agenda for a teambuilding session is shown in Figure 3–6. The times shown are just guidelines developed from experience. Wide variations in the time required for sessions is common. The agenda calls for six sessions:

Session 1: *Introduction and background* on why the teambuilding training is taking place, including goals and objectives. This information should be presented by a senior executive.

Session 2: *Description of the teambuilding training* that will take place. This session is presented by the facilitator.

Session 3: *Low-risk interaction*, such as listing and discussing team members' concerns for the business and the things about the business of which they are proud. These should be listed in a free-flowing session using brainstorming rules. Participants should reach consensus on the key concerns and points of pride.

Session 4: *Discussion of teamwork fundamentals*, including the need for trust, respect, and mutual support. This session is presented by the facilitator.

Session 5: *Medium-risk interaction* on what participants perceive as strengths that will help the group succeed as a team, as well as any impediments that threaten success. These strengths and impediments should include both interpersonal and organizational factors. Once expressed, these factors should be discussed and resolved. Nothing should be left hanging. At the least, an action item with definite follow-up and a suspense date should be established. The facilitator must prevent this session from becoming personal, so as to ensure an atmosphere of mutual support and respect.

Figure 3–6
Outline of Typical Team
Building Training Session

Outline TEAM BUILDING TRAINING		
Session Number	**Session Topic/ Issue**	**Approximate Time**
1.	The WHY of Teambuilding	1 Hour
2.	Description of the Training that Will Take Place	1/2 Hour
3.	Identification of the Organization's Strengths	2 Hours
4.	Teamwork Fundamentals, Trust, Respect, Mutual Support	1 Hour
5.	Identification of Impediments to Teamwork	4 Hours
6.	Development of an Action Plan for Overcoming the Impediments	4 Hours

Session 6: *Discussion of actions* that can be taken to overcome factors that inhibit effective teamwork. Consensus is the objective here. This session is led by the facilitator. Problems, solutions, and action items are recorded.

These six training sessions usually take approximately twelve hours. Time should be built in for breaks at least every two hours and for meals at appropriate times. It is also important to schedule group-oriented fun time for participants. This is not merely for relaxation. If properly conducted, such activities can contribute to teambuilding. If time permits, team-success activities should be provided. These activities can include solving one or more of the problems that have been identified, or developing a plan to do so. Teambuilding can easily require two or even three days.

SOME CAUTIONS

Teambuilding is a process that can yield great rewards in terms of improving the group's ability to get things done effectively and efficiently. Nevertheless, there is a potential downside that must be guarded against, and precautions must be taken if the teambuilding effort is to be a success.

Prevent Alienation of Team Members

There is always the risk of overstepping the bounds of teambuilding, resulting in the alienation of one or more of the team members. Teambuilding is, to a large extent, about building personal relationships. Any time personal relationships are involved, there is always the risk of alienation. The leader and facilitator must be aware of this danger, and be vigilant to prevent it. The following strategies will help ensure that personal animosities don't develop and lead to alienation of one or more team members:

Make sure communication is complete.   The facilitator should clarify and test understanding by paraphrasing and restating comments made by participants. For example, in the course of a teambuilding session, Bob says to Bill, "Getting anything out of your department is just too hard." Does this comment mean that Bill's department is slow, incompetent, uncooperative, or what? On the face of it, the comment seems to mean that Bill's department is difficult to deal with, but it is important that everyone understand precisely what is being said. The facilitator should seek clarification. He or she might ask Bob if he means that Bill and his personnel lack a can-do attitude. By the time the discussion runs its course, the answer could very well be that Bill's department is so bound up by outdated procedures that they have difficulty doing their job. They are probably more frustrated than Bob is. In this case, a problem that seemed to be personal turned out to be organizational. This won't always be the case, of course. But even when it isn't, the message should be fully and clearly understood. Invite all participants to air personal issues.

Don't avoid negative comments.   The insights team members provide may not have occurred to the offenders. In addition, such comments suggest areas of behavior that might be improved. The facilitator's job is to prevent the discussion from deteriorating into one that alienates participants. When a personal issue is raised, look for and suggest ways to overcome the problem that allow both parties to benefit.

Don't let a participant become defensive.   Once this happens, it can be difficult to break through the defensive paradigm. Be prepared to be seen as part of the problem. Should the discussion indicate that the leader is part of the problem, he or she must be especially careful to accept the comments, to gently probe for more information, and to refrain from becoming defensive or irritated. Facilitators can lighten the blow by telling the leader to expect such comments, and that they are normal.

Develop Trust among Team Members

Trust is another major issue in teambuilding. Developing trust among a team's members is one of the primary objectives of teambuilding. Cautions associated with developing trust are as follows:

- If an organization has had problems with trust, it is not likely that a teambuilding session will completely eliminate them. Trust must be earned, and can be achieved only by observing trustworthy behavior over an extended period of time. Do not expect a quick fix when working to establish trust.

- One of the points stressed in a teambuilding process is that *what is said in a training session should go no further.* All team members must observe confidentiality. This is essential if participants are going to open up. If comments which were made with the expectation of confidentiality later surface in the organization, it will be clear that someone has breached the trust, and any gains made in teambuilding will be wiped out, perhaps forever.

■ When a team leader is informed that he or she is part of the problem, the relationship with the participant making the comment must not change or be allowed to suffer. Teambuilding requires frank, open discussion. A team cannot withstand retribution without losing trust.

Be Scrupulously Honest

For a team to function effectively, honesty is an absolute requirement. Many organizations are so byzantine in their internal politics that they seem to exist for intrigue rather than for any business or social motives. It will be difficult for people who have always worked under these conditions to suddenly become open and honest. If a leader presides over an organization in which honesty is not taken for granted, he or she will have to set an example of being open and honest and stick to it over time before team members accept his or her leadership. As with trust, establishing honesty is not a simple matter of saying, "From now on we're going to be open and honest with each other." Time can heal, but only if the concepts of openness and honesty are always meticulously observed. Be sensitive to the fact that even an expedient falsehood, no matter how trivial or well-intended, can erase all gains in earning a reputation for honesty. In a Total Quality setting, there is never sufficient reason to lie to an employee.

The focus of attention has been on the leader in this section, but what applies to the team leader also applies to all team members and employees.

Be Knowledgeable and Well Prepared

The leader of the team should be well prepared for the teambuilding process. Team leaders should prepare and distribute an agenda, handouts describing the various activities planned, and the background and objective of teambuilding. He or she should do this even if an outside facilitator will be used. The leader must appear knowledgeable about teambuilding and the need for and expectations of the process. The team leader must be perceived as a knowledgeable, committed advocate of teambuilding. The leader must be sufficiently knowledgeable of teambuilding and related subject matter to *lead* discussions. If the facilitator has to assume that role, it will be at the expense of the leader's credibility. (And remember, the idea is to lead, not dominate.)

Teambuilding gone wrong can have disastrous results. This is why it is essential for the leader to objectively consider whether outside help is needed in facilitating the sessions. If the team leader is not adequately prepared or equipped to facilitate teambuilding activities, but proceeds without outside help, the organization will be set back to a point that is worse off than when it started.

TOTAL QUALITY TIP

Team Building Requires Cultural Change

"A total quality commitment brings change to the entire organization and its culture. Certain systems, priorities, and procedures have to be adopted right away so that the organization is ready for these changes."[2]

Raymond T. Bedwell, Jr.

Maintain Executive Commitment

As with all changes to an organization's culture, teambuilding requires ongoing commitment from the top. The following concepts apply to maintaining a commitment from the top:

- Often the extent of executive commitment is proportionate to the size of the crisis. When the crisis passes, so does the commitment. This is akin to giving up a diet and exercise program once the weight has been lost. Just because things are going reasonably well doesn't mean they cannot be improved. Without a continuous commitment, teambuilding will not succeed. Success over time requires that the leader be directly, actively, and personally involved, and involved permanently, not just for the few days of formal teambuilding activities. If the leader cannot make the commitment, it is better not to start the effort.

- Permanent commitment requires that the team leader change his or her behavior appropriately, and sustain the new behavior pattern over time. If he or she is not committed, this is not likely to happen, and the rest of the team will correctly interpret the situation as the end of teambuilding . . . and probably of Total Quality.

- The leader must reward and reinforce positive behavior in the team and in the organization-at-large on a continuing basis. Again, this is not likely to occur unless there is a full and unwavering commitment, because reinforcing behavior is not something that can be taken lightly. Failure to follow through can put an end to teambuilding and endanger total quality throughout the organization.

- Just as positive behavior must be continually reinforced and rewarded, negative behavior must be eliminated, ultimately by removing the guilty parties if less drastic measures do not achieve the desired results. It takes long-term commitment to ensure that negative behavior by a few doesn't destroy the efforts of many.

- During the formal teambuilding sessions, a list of problems for resolution and a corresponding list of action items will be developed. Management will need to make a concerted effort to ensure that these action items are pursued and executed. This, too, requires long-term commitment. Do not relax the effort at the end of the formal teambuilding activities.

Formal teambuilding sessions represent only the beginning of real teambuilding. It is important that team members know that the job is not finished at the end of two or three days of formal training. Remember, as one set of problems is dispensed with, there will always be another set to replace them. These, too, must be dealt with, or teambuilding will cease and the team's potential will never be fully realized. This is the team equivalent of continuous improvement. Never assume that there are no more problems or opportunities for improvement.

THE NEED FOR OUTSIDE ASSISTANCE

It has been shown that the leader must decide whether to act as the facilitator of teambuilding activities or bring in outside help. Team leaders should ask themselves the following questions when considering this decision:

- Are you certain that your fellow team members do not perceive your behavior, attitudes, or values to be inhibitors of teambuilding?
- Is the team free of interpersonal issues that might hinder teambuilding? Are there issues perceived by team members as having been ignored by you?
- Would your team be more comfortable with you than with an outsider in the facilitator's role?
- Are you confident that you can get the team to open up and be completely honest?
- Will you be able to defuse personal issues and defensive reactions in a manner that prevents damage to the teambuilding effort?
- Are you thoroughly grounded in the teambuilding process?

Having considered these questions objectively, if all of a team leader's answers are yes, he or she should be able to conduct the team building sessions without outside help. However, unless all six questions can be objectively answered yes, an outside facilitator should be brought in for the training sessions.

FOLLOW-UP

The initial formal sessions are not the end of teambuilding. They are just the beginning. From this point forward, the process should be continuous, and both the leader and team members should be sufficiently familiar with the process that outside help is no longer required. However, it is easy to slip back into old, familiar patterns, and the team leader must stay alert to prevent this from happening. Following are several strategies that can help the team leader and team members.

- At every team meeting, review the list of action items developed during the formal sessions, and make sure that real progress is being made. Insufficient time can no longer be an excuse, because it indicates that nothing has changed, and time is still being spent reactively (i.e., putting out fires).

- At every opportunity, improved behavior should be rewarded and reinforced. Similarly, every example of improper behavior should be corrected immediately.

- Solutions that were developed during the formal sessions, or those that spring from the action items, will require nurturing until they mature and become accepted as standard practice. Consequently, solutions should be monitored closely to ensure that they become normal operating procedure.

- Team successes should be celebrated. Team failures should be analyzed and learned from. Never let analyzing a failure turn into, or be perceived as, looking for someone to blame.

The team should have as one of its objectives the continuous improvement of team effectiveness. All team members should be conscious of the need to become more team oriented every day, now and forever. When team members no longer think in terms of individual departments, teamwork has begun to take root. When difficulty in one department becomes of personal concern to every team member, real progress has been made. But like the continuous improvement of any process or product, there will never be a time when the teambuilding effort is complete. There will always be room for improvement, especially as people leave and others join the team.

SUMMARY

1. Teambuilding means identifying and solving interpersonal, organizational, and procedural problems within a team, thereby making it possible for individual members to work cohesively as a team.

2. The goal of teambuilding is to enable team members to work together toward a common objective, complementing each other with their skills and experience and offsetting strengths and weaknesses.

3. Teambuilding can help any organization, even those that are working well. It is particularly needed in organizations that are experiencing turnover in higher management, excessive work hours by higher managers, quality problems, excessive waste, conflict, interdepartmental barriers, poor communication, morale problems, and customer dissatisfaction.

4. In order for teambuilding to succeed, an open constructive, nonthreatening environment must be maintained. Team members should be mutually supportive so that trust can develop.

5. The teambuilding process revolves around four questions: Who? Where? When? What? The agenda for training sessions should include the following topics: Intro-

duction and Brainstorming; Description of Team Building; Identifying Strengths, Teamwork Fundamentals; Identifying Inhibitors, and Developing an Action Plan.

6. To avoid alienation of team members, make sure communication is complete, don't avoid negative comments, and don't let participants become defensive.

7. Trust and honesty are critical in teambuilding. Trust must be built over time by providing an ongoing example of honesty and confidentiality.

8. Permanent commitment is critical. Just because things begin to work reasonably well doesn't mean that they cannot be improved even more. Permanent commitment means that the team leader has to change his or her behavior appropriately and sustain the new behavior pattern permanently.

9. An outside facilitator is needed when the team leader might be viewed as an inhibitor; when interpersonal issues among team members have been ignored in the past; or when team members would be more comfortable with an outsider.

=== **KEY TERMS AND CONCEPTS** ===

Customer dissatisfaction	Interpersonal problems
Environmental considerations	Low-risk interaction
Executive commitment	Medium-risk interaction
Facilitator	Organizational problems
Follow-up	Procedural problems
Honesty	Quality problems
Individual achievement	Team building
Interdepartmental barriers	Trust
Internal competition	

=== **REVIEW QUESTIONS** ===

1. Define the term *teambuilding*.
2. What is the goal of teambuilding?
3. Explain the benefits of teambuilding.
4. Describe the ideal environment for teambuilding.
5. Summarize the responsibilities of the following participants in teambuilding: team leader, team members, facilitator.
6. Explain the *Who, What, When,* and *Where* of teambuilding.
7. Describe a typical agenda for teambuilding training.
8. How can the potential downside of teambuilding be avoided?
9. Explain how to maintain executive commitment to teambuilding.

ENDNOTES

1. Robert M. Tomasko, *Rethinking the Corporation: The Architecture of Change* (New York: American Management Association 1993), 91.
2. Raymond T. Bedwell, Jr., "How to Adopt Total Quality Management: Laying a Sound Foundation." *Nonprofit World,* Volume II, Number 4, July/August 1993, 28.

CASE STUDY 3–1

Building the Steering Committee Team at MTC

The MTC Steering Committee members had decided that the team leader, John Lee, should double as the facilitator during teambuilding training activities. Lee was pleased with their confidence in him, but he had strong reservations. Since he had put the authors of *Introduction to Total Quality* on retainer as implementation coaches, he asked them to conduct a brainstorming session with the Steering Committee to identify the advantages and disadvantages of having the team leader double as teambuilding training facilitator.

The major advantage of having Lee serve as facilitator was his positive relationship with each member of the Steering Committee. However, Lee's desire to be a full participant in the training, coupled with his doubts about having sufficient facilitation skills, convinced team members that an outside facilitator was appropriate after all.

One of the implementation coaches agreed to serve as the teambuilding training facilitator, and plans for the training began immediately. The Steering Committee decided on two intensive full-day sessions at a local resort. Team members were to check in at the resort on the following Monday morning. Training was scheduled to begin promptly after a group breakfast and conclude at 4:30 p.m. the following day.

The activities proceeded without incident until Session Four. During this session, two things happened simultaneously. First, several of the participants became defensive when certain of their behavior patterns were singled out as impediments to teambuilding. When they became defensive, John Lee responded by becoming protective of them. The facilitator pointed out how both types of behavior would impede teambuilding and that frank, open, non-judgmental discourse was a difficult but necessary aspect of teambuilding.

It took a great deal of effort on the part of all participants and vigilance on the part of the facilitator, but team members eventually stepped out of their defensive, protective paradigms and identified some inhibitors that had to be overcome. Inhibiting behaviors exhibited by various team members included: a reluctance to accept responsibility when something goes wrong; a tendency to claim credit even when no credit is due; an overly aggressive confrontational approach when dealing with even the most insignificant issues; and a tendency to seek an inordinate amount of the boss's time.

In the final session of the second day, a plan was developed for overcoming the inhibitors identified. The facilitator had planted the seeds of a teamwork paradigm. It would now be the job of the team leader and the other team members to continue developing as a team. They all agreed that teambuilding had just started, but that this had been a good start.

=========== **CASE STUDY 3–2** ===

Teambuilding Problems at ESC

Lane Watkins had had a hard day. His assignment to implement Total Quality wasn't going very well. Sure, he had a Steering Committee to work with, but there wasn't even one executive-level manager on it. The middle managers to whom this responsibility had been delegated were good people, but they didn't have the level of authority needed. To make matters worse, they weren't sure how much support they had from top management. Executive commitment had been proclaimed by John Hartford, but beyond his initial pep talk there had been very little evidence of commitment.

Teambuilding had been a disaster. First, John Hartford had refused to approve the recommendation for an outside facilitator. Then he had refused to allow Watkins to arrange an off-site location for the training. To make matters worse, two members of the Steering Committee were not allowed to participate and one had to leave early. Finally, Watkins was required to conduct the training himself.

Predictably, things didn't go very well. After two hours of trying to keep things on track, Watkins had finally thrown up his hands in frustration. The agenda went out the window and what was supposed to have been a training activity dissolved into a gripe session. One of the team members spoke for the rest when she said, "What's the point here? Even if we become the best team in the world, which we obviously won't, we don't have the authority to make high-level decisions. No matter how well we work together, it won't matter. I suggest we go back to work and tell anyone who asks that the training went just fine."

The next morning Watkins was scheduled to meet with John Hartford to give him a progress report. Initially Watkins had hoped this assignment would result in a promotion. "How quickly things change," thought Watkins. "With the way the things are going, I'll be lucky just to keep my job."

Train the Steering Committee in the Fundamentals of Total Quality

At this point, top management has committed itself to Total Quality, the Steering Committee has been formed, and teambuilding has begun. Some of the Steering Committee members may understand the fundamentals of Total Quality, but it is usually safe to assume that most do not. The Steering Committee is about to undertake its role of leadership in the implementation of Total Quality and in managing the organization according to its principles. It follows, then, that the team needs a strong foundation in the fundamentals of Total Quality.

Total Quality training for the Steering Committee typically takes two to four days. Like the teambuilding sessions discussed in Step 3, this training must include the entire Steering Committee, full time, for the duration of the training. Consequently, training is best done off-site, so that there are no interruptions or disturbances, and in relaxed, casual surroundings that promote attention and interaction. The training days should be scheduled to include eight hours of classroom sessions broken up by breaks and lunch. Informal discussion should be planned for the evenings.

RATIONALE FOR TRAINING THE STEERING COMMITTEE

The rationale for training the Steering Committee is illustrated in Figure 4–1 and explained in the following paragraphs.

Figure 4–1
Why Train the Steering
Committee?

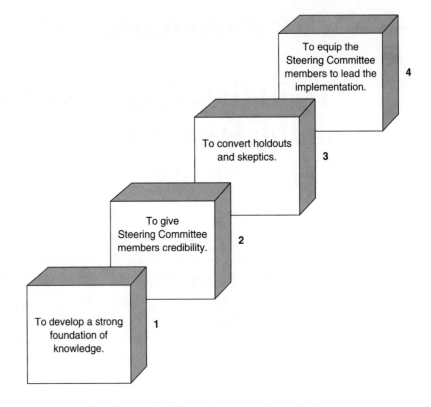

- *To develop a strong foundation of knowledge.*   Since the Steering Committee will guide the entire organization in the implementation of Total Quality, it is important that Steering Committee members be well versed in the fundamentals of Total Quality.

- *To give Steering Committee members credibility.*   The implementation of Total Quality means organization-wide change, so it is critical that Steering Committee members have credibility. Members who can speak the language and apply the principles of Total Quality will have credibility. They develop the ability to do these things during their training sessions.

- *To convert the holdouts and skeptics.*   Some of the Steering Committee members may not be sold on the Total Quality concept at the outset. Training in the fundamentals of Total Quality is a way to convert them, or at least to eliminate any outward hostility they may retain due to misconceptions they may have about what Total Quality will entail.

- *To equip the Steering Committee members to lead implementation.*   At the conclusion of the training, the Steering Committee should be well equipped to begin its work leading the organization in the implementation of Total Quality.

TOTAL QUALITY TIP

Training the Steering Committee Is Crucial

"Training must take place at all levels within the organization. However, it must start at the top. Management must be the driving force behind the transformation process. Everything can improve—even management."[1]

Stephen Uselac

WHO SHOULD DO THE TRAINING?

Training should be provided *by someone who has been there*. In a practical sense, this usually means that training in fundamentals must be done by an outside consultant or consulting team. There are exceptions to this, of course. For example, if the organization is a division of a larger corporation with other divisions that have successfully implemented Total Quality, it may be possible to draw expertise from the other divisions. Most often, however, it will be necessary to retain an outside consultant. Providing Total Quality training should *not* be undertaken by the team leader unless he or she has handled a successful implementation in another organization.

THE TRAINING CURRICULUM

A basic curriculum which covers the fundamentals of Total Quality is given in Figure 4–2. The curriculum recommended is one that will provide a fundamental understanding of the history, philosophy, rationale, and working tools of Total Quality. To discuss all the course material in this chapter would require more space than is available. Consequently, the companion text to this book, *Introduction to Total Quality*,[2] is recommended as a single source from which to draw the course material. The intent of the discussion that follows is to provide a brief summary of the topics to be covered.

The History of Total Quality

It is impossible to develop an appreciation for the power and wisdom of Total Quality without having an understanding of its history and development. Many otherwise well-informed people mistakenly believe that Total Quality simply involves getting back to the basics, or that it is just plain common sense. The obvious response to comments of this nature is, "If Total Quality is just common sense or just good business basics, then why haven't you been applying its principles in your organization up to now?" The fact is, Total Quality does make sense to most people who acquire an understanding of it, and it does promote doing things in simpler ways. But it is not just the application of common

Figure 4–2
Basic Curriculum on
Fundamentals of Total Quality

<table>
<tr><td colspan="2" align="center">Basic Currriculum
FUNDAMENTALS OF TOTAL QUALITY</td></tr>
<tr><td colspan="2">Training for the Steering Committee and other groups unfamiliar with the Total Quality approach should cover the following topics as a minimum:</td></tr>
<tr><td>☑</td><td>The History of Total Quality</td></tr>
<tr><td>☑</td><td>The Total Quality Philosophy</td></tr>
<tr><td>☑</td><td>The Rationale for Total Quality</td></tr>
<tr><td>☑</td><td>Changing the Organization's Culture</td></tr>
<tr><td>☑</td><td>Statistical Tools for Total Quality</td></tr>
<tr><td>☑</td><td>Implementing Total Quality</td></tr>
<tr><td>☑</td><td>The Role of the Organization's Vision</td></tr>
<tr><td>☑</td><td>Planning the Implementation</td></tr>
</table>

sense, nor is it a simplification of traditional systems of doing work. It is, in fact, a highly developed way of doing business that is unlike anything that has been done before.

In order for trainees to fully appreciate the contrast between Total Quality and traditional ways of doing business, the training must cover traditional management as it developed from the craft shops of the nineteenth century, to the mass production and specialization of the early and mid-twentieth century, and to the Total Quality approach of the late twentieth century. In each case, the strengths and weaknesses of the methods employed should be pointed out and discussed thoroughly. In addition, the people who pioneered the concept of Total Quality should be discussed, including the following:

Frederick W. Taylor	For the scientific management theory.[3]
Henry Ford	For mass production and perfect interchangeability of both parts and workers (i.e., division of labor).
Walter A. Shewhart	For developing the concepts of applying statistics to quality improvement.
W. Edwards Deming	For carrying Shewhart's torch in finding a receptive audience in Japan and broadening the Total Quality philosophy.
Joseph M. Juran	For expanding the Total Quality knowledge base.
Kaoru Ishikawa	For developing Total Quality Management.
Taiichi Ohno	For developing Just-in-Time production.
Genichi Taguchi	For developing advanced statistical tools.

Another important aspect of Total Quality that should be noted is the role a crisis environment played in its acceptance. The first widespread use of Shewhart's techniques came about in the United States because of the crisis brought about by the need to produce military equipment during World War II. The second wave in the development of Total Quality came in Japan in the early 1950s as that nation, in economic crisis as a result of its defeat in World War II, found that it had to improve the quality of its products in order to compete in world markets. Japan adopted the teachings of Deming and Juran and began to expand and develop their concepts. The third wave of development began in the United States in the late 1980s as more and more manufacturers faced the crisis of decreasing market share as the result of intense foreign competition.

Crisis has historically preceded the beginnings of all Total Quality implementations. Crisis is a powerful motivator that can cause organizations to abandon failed procedures and cultures and try something new. If things are going well, it is difficult for an organization to muster the will, the energy, and the stamina to change. However, when an organization is faced with a crisis, reluctance to change can usually be overcome.

There is a second element of the environment created by a crisis that plays a role in the acceptance of Total Quality: consumers' heightened expectations concerning the quality of products or services. Consumers have learned to demand quality. Therefore, they are becoming less and less willing to settle for second-rate goods or services. This is the factor that will eventually eliminate organizations that refuse to change, or that are too late in accepting the need for change.

Another factor that encourages adoption of Total Quality is that society is becoming increasingly concerned over the waste of precious resources, both natural and human. Examples of intolerable waste are inefficient processes that waste material and other resources; products that do not work as they are supposed to, that have short life spans, or that cost too much; and services that must be repeated again and again before the intended effect is achieved. Such waste is no longer acceptable to consumers who know that high quality is not just possible, but available from an increasing number of Total Quality producers and service providers.

The Total Quality Philosophy

Total Quality is important not just as a management method, but as a management philosophy. Kaoru Ishikawa defined Total Quality as a *thought revolution in management*.[4] According to Dr. Ishikawa:

> If TQC is implemented company-wide, it can contribute to the improvement of a company's corporate health and character. TQC is one of the major objectives of the company. It is its new management philosophy. Set your eyes on long-term profits and put quality first. Destroy sectionalism. TQC is management with facts. TQC is management based on respect for humanity. TQC is a discipline that combines knowledge with action.[5]

The following paragraphs describe the elements of the Total Quality philosophy that should be covered in the training session.

Quality First

Most Western business organizations are still obsessed with short-term profits. Total Quality changes this obsession to one of *quality first*. Once this is done, long-term profits will follow.

Customer Orientation

The focus of the organization must be on its customers, not on itself. In the final analysis, it is the customer, not the organization's management, who decides whether an organization is successful. Customer orientation must also be applied to the organization's *internal customers*.

Management by Facts

Too many organizations are managed by the so-called intuition of their managers. There probably never was a time when management by intuition was appropriate, but it is clearly inappropriate in this age of global competition. Management decisions at all levels must be based on facts, not on emotion or preconceived notions. The typical organization today has no shortage of data. The key is to sort out what is important and convert data to facts that have immediate application.

Employee Involvement and Empowerment

Most employees want to do a good job. In addition, most employees have ideas about how to do their work better, faster, more accurately, or more consistently. Unfortunately, there are too few situations where employees are fully empowered to do a good job and to turn their ideas into improvements. This is one of the most difficult aspects of the Total Quality philosophy for many managers, yet it is absolutely essential to a successful implementation. Managers must learn that they have no monopoly on wisdom. They must respect their employees and acknowledge that every employee has knowledge that can and should be tapped. An axiom of Total Quality is that the people best able to improve a process are those who work with that process day in and day out: the employees. Managers should take this axiom to heart.

Building Trust

In a Total Quality organization, trust among all parties is essential. Managers must develop trust in their employees, and employees in their managers. This requires trustworthy behavior on the part of the managers, and the opportunity to demonstrate trustworthiness on the part of employees. Ethics, integrity, and character must replace politics, intrigue, and backbiting. While trust is necessary, it cannot be mandated; it can only be earned. Some organizations will start with a considerable deficit in their trust account. Others that have begun building trust will find that their progress can be set back by a single misstep. Building trust is one of the most important tasks the Steering Committee can undertake.

Relentless Pursuit of Continuous Improvement

Continuous improvement of all its products and services, both for external and internal customers, is a top priority of the Total Quality organization, and it is one that never stops. A sense of having made all the improvements possible never develops. Instead, the organization and all those in it must believe that there is always room for more improvement. The organization should recognize that there are two kinds of improvement. The first is *incremental improvement* and is the one in which most people are involved. The second is *sudden and dramatic advances*. These are usually confined to the few processes or products that were far inferior to similar processes or products in competing organizations. Incremental change is typified by benchmarking. The logic and imperative for relentless pursuit of continual improvement are easy to embrace, but the discipline necessary to accomplish this is more difficult. Constant attention and encouragement by the Steering Committee are required.

Total Organization Involvement

Total Quality cannot be achieved in one or two departments while the other departments go on with business as usual. By definition, Total Quality means all functions are involved. This is because in a Total Quality setting, departments no longer operate as stand-alone units, but are cross-functionally managed by the Steering Committee. Engineering departments are no longer independent of manufacturing input. Accounting departments evolve into true service centers for the organization. In fact, departmental walls eventually become transparent. The organization may find it difficult to deal with this philosophical paradigm at first, but it is fundamental to Total Quality. Another imperative in a Total Quality setting is that all processes used by the organization must be subjected to continuous improvement. For example, it is not sufficient to confine process improvement to manufacturing. All processes are targets for improvement.

Teamwork Rather Than Individual Achievement

The traditional approach to doing work has been individual achievement. In a Total Quality organization, most of the work is done by teams. The team may be a natural work group, as in the case of a production cell, or a cross-functional group whose members support each other with offsetting strengths and weaknesses. There is no attitude of "I've done my part, now it's all yours." Often teams, particularly those established to tackle specific problems, are made up of members from different departments and with different backgrounds. For example, if the team is to solve a process problem, it should include at least the person who operates the process, the person who supplies the process, and internal customers of the process. The key is to bring as much brain power as possible and as many points of view as possible to bear on the problem.

Continual Opportunities for Learning

Affording all employees opportunities for continual learning is an important element of Total Quality. More is expected of employees in a Total Quality setting, so it follows that training will be required. The organization should budget a percentage of annual earnings or a certain number hours per employee per year for training. Such an investment will pay dividends in terms of improved employee performance.

Unity of Purpose

Although unity of purpose is listed last here, it is by no means least in importance. Total Quality organizations ensure that all employees, regardless of position, are pulling in the same direction toward a common goal. Internal competition, with its wasted energy and resources, is discouraged. Departmental or personal objectives that are at cross-purposes with the larger objectives of the organization are also discouraged. As simple and logical as unity of purpose sounds, the fact remains that few traditional organizations operate this way. Few employees have a thorough understanding of their organization's objectives to guide their efforts. The issue here is communication, or the lack of it. Total Quality organizations have a vision and a set of operating principles that are continually communicated throughout the organization. They have a set of goals and objectives supporting their vision, developed with input from the entire organization and kept in front of all employees continually. Progress toward accomplishing the organization's goals and objectives is reported regularly to all employees. Even the financial results are shared with employees.

Effective publicizing of the vision, operating principles, objectives, and performance must be the rule. There should be no secrets, no withholding of information. Fully informed employees will have the information they need to work better and smarter, and to align their efforts with the objectives of the organization. Many Total Quality organizations go beyond informing employees and also include customers, suppliers, stockholders, the communities in which the organization resides, and anyone else who has a stake in the success of the organization.

The philosophy of Total Quality owes much to Dr. W. Edwards Deming. Consequently, Deming's Fourteen Points and his Seven Deadly Diseases, discussed in the Introduction and presented in Figures I–5 and I–6, should be covered in this section of the training curriculum.

The Rationale for Total Quality

The simplest rationale for Total Quality is competitiveness. An organization can either adopt this approach or go out of business. This statement may seem overstated to business people who have avoided a crisis of competition to this point. Nevertheless, the fact is that organizations that have embraced Total Quality are gaining market share against those that have not. It is just a matter of time before the rationale stated above is proved. It has already been proved for many organizations—and for entire industrial sectors, in fact. North America once had a thriving consumer-electronics industry. Today, this industry is virtually nonexistent, having been lost to foreign companies that make better products at competitive prices. The American automobile industry has lost more than 30 percent of its market share to the foreign competition for the same reason. In the United States, General Motors is struggling not only against its foreign competitors, but also

---

TOTAL QUALITY TIP

There Must Be a Commitment to Change

"Commitment to change occurs when managers become aware of the discrepancy between the newly adopted theory and their own everyday behavior. However, this commitment develops only if the managers believe that they will benefit personally by changing their style of management. Initial attempts are often discouraging and if not reinforced by some type of rewarding feedback, may gradually be discontinued. Commitment and reinforcement must be strong and continuous to overcome established patterns. Moveover, behavioral changes are often viewed with suspicion by people whose opinions of the manager were formed by long exposure to his or her previous style."[6]

M. Scott Myers

---

against domestic competitors that have adopted Total Quality. There is a message here, even for organizations that have no foreign competitors. Their survival can be threatened just as much by enlightened domestic competitors that adopt the Total Quality approach. There is no safe place to hide. Total Quality is the most competitive approach to doing business. Therefore, the sooner it is adopted and implemented, the better.

One should not assume that Total Quality applies only to the world of business. The educational system, government at all levels, and the military are also under tremendous pressure to do more with less. The public sector is facing challenges to provide services at lower costs and maintain a strong national defense in the face of drastic funding cuts. Since the public sector is often less efficient than the private sector, it has even more to gain from adopting the Total Quality philosophy.

Changing the Organization's Culture

Every organization has its own unique culture. In a Total Quality context, the term *culture* means the concepts, habits, expectations, and accepted behavior of an organization. An organization's culture is a nonverbal way of saying, "This is how we do things here." Steering Committee members should understand that the current culture of their organization may be a roadblock to the implementation of Total Quality. An organization's culture is established over time and reinforced continually by both organizational policies and management practices. Consequently, it can be difficult to modify the behavior of an organization. Trying to do so is a lot like trying to change individual behavior, as when one tries to stop smoking or lose weight. Regardless of the difficulty, however, changing the culture is necessary. Steering Committee members should also understand that it is their responsibility to change the culture because they are the ones who create the organization's culture, good or bad.

In attempting to change the organization's culture to embrace the Total Quality philosophy, actions speak much louder than words. Changing the culture cannot be done with orders or pronouncements, but can only be accomplished by actions that are in keeping with the desired culture. Words can be used to explain, but it is the actions of managers and training that lead to cultural changes. The commitment and all related actions must be consistent. Management cannot be for quality first today and change its mind tomorrow. Dr. Joseph M. Juran suggests the following strategies for introducing cultural change:[7]

Involve everyone.   All members of the organization should participate in planning and executing the change.

Give it time.   Allow time for members of the organization to determine what the changes will mean to them. Also allow time to identify and involve the advocates of change.

Start small.   It is best to start small rather than try to accomplish the cultural change in a single step. This approach reduces risk and gives people time to evaluate what is happening.

Choose the right time.   Juran actually says, "Choose the right year," but the point is, be sure the timing is right. For example, an organization that is preparing to undergo some other substantial change should not introduce cultural change at the same time.

Keep proposals for change well focused.   Keep a clear focus on the desired change. Do not allow extraneous matters to cloud the issue and cause a loss of focus.

Convince the cultural leadership.   If the leaders of the culture—formal and informal—are convinced of the benefits of change, they will play a leadership role in convincing the rest of the organization.

Treat all employees with dignity and respect.   People who are treated with respect and dignity are more apt to accept the change than employees who are simply told, "This is the way it is going to be."

Look at the change from the other person's perspective.   Try to understand the impact of cultural change on the individuals involved. This will help managers anticipate potential problems and reactions.

Consider alternatives.   There is always more than one way to do something. In dealing with resistance to change, there are a number of alternatives, including persuasion, developing win-win strategies, altering change proposals to meet specific objections, and even abandoning the change if necessary.

Statistical Tools

Steering Committee members will require enough hands-on training in the use of Total Quality tools to be able to employ them in problem solving, and perhaps to instruct subordinates in their use. The Total Quality tools are described briefly in the following list, and several are illustrated in Figure 4-3. For a complete explanation of how to use these tools, refer to the companion book, *Introduction to Total Quality*.[8]

Pareto Charts, to separate the important from the unimportant.

Fishbone Diagrams (also called Ishikawa Diagrams and Cause-and-Effect Diagrams), to identify the possible causes of a problem and determine those likely to be responsible.

Stratification, for grouping data into categories in order to isolate root causes.

Check Sheets, to facilitate the collection and interpretation of data for specific purposes.

Histograms, to display the frequency distribution of data and facilitate its interpretation.

Scatter Diagrams, to determine the correlation between two characteristics.

Run Charts and Control Charts, to record the output of a process over time (run charts) and to control and improve processes (control charts). These are the instruments of Statistical Process Control.

Flow Charts, to diagram processes so they can be more fully understood and to identify possibilities for improvement.

Surveys, for the acquisition of data.

Plan-Do-Check-Act Cycle (also known as the Shewhart Cycle or the Deming Cycle), by which plans are developed and put into action. As the action is taking place, feedback is received by the planners, who make any necessary changes to the planned action, and the cycle repeats. This tool provides a continuous means of fine-tuning and course correction.

These tools will be as useful to Steering Committee members as the saw and hammer are to carpenters. As the Total Quality process matures in the organization, additional, more sophisticated tools may be introduced, but these are adequate for the early stages.

Implementing Total Quality

The Steering Committee must be acquainted with the different approaches to implementation. There is no one right way to implement Total Quality. However, there are certain fundamental rules that should be observed, and there are specific steps to be carried out in sequence over time that will yield a high probability of success. Figure I–12 in the Introduction to this book illustrates the implementation steps in three distinct phases: Preparation, Planning, and Execution. The figure also shows who is responsible for each step. Figure I–12 should serve as the central focus of this component of the training.

The important concept for the Steering Committee to grasp at this point is that the implementation should be structured and well planned, carried out in a sequence of small steps, and backed by unwavering commitment from the top. It is also important for the Steering Committee to recognize that many of the steps, once executed, never stop. Each of the twenty steps in the implementation process is represented by a chapter of this book; therefore, no attempt is made to summarize the process here.

The Role of the Organization's Vision

Total Quality requires a long-range vision because it takes time to fully implement. It involves fundamental changes in the way organization's do things and how people work

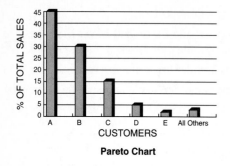

Pareto Chart

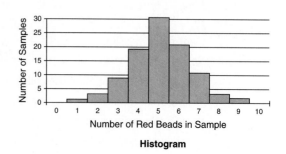

Histogram

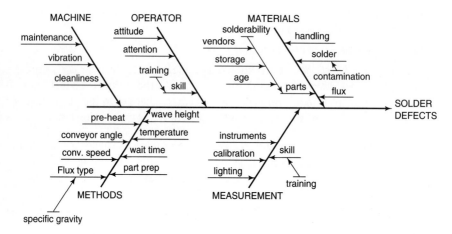

Fishbone Diagram

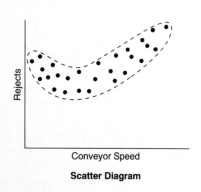

Scatter Diagram

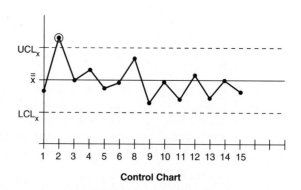

Control Chart

Figure 4–3
Total Quality Tools

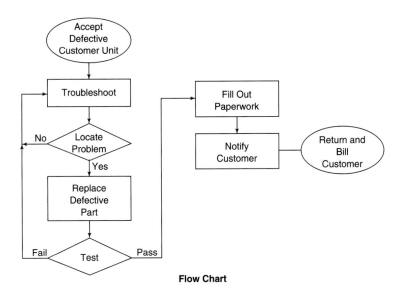

Flow Chart

Check Sheet

Shaft length - Week of <u>7/11/94</u>   (Spec: 1.120 - 1.130")

- -

1.118**	13
1.119**	11
1.120	11 13
1.121	12 13 15 15
1.122	11 11 13 14 14 15 15 15
1.123	11 11 11 13 13 13 14 15 15 15
1.124	11 12 12 12 13 13 14 14 14 15 15 15
1.125	11 12 12 12 12 13 13 14 14
1.126	11 12 12 13 14 14
1.127	12 12 14 15
1.128	11 11
1.129	14
1.130	
1.131**	
1.132**	

** Out of Limits (at rows 1.118 and 1.119)

Enter day of month for data point. (at rows 1.129–1.130)

- -

Check Sheet

together, and involves customers and suppliers in ways never before considered. It also places value on matters that may never have been discussed. Total Quality does not come together overnight. Therefore, the organization's vision must be long enough in scope to provide a consistent course as the implementation moves forward. Without a long-term vision, the organization will find itself constantly diverted by minor detours. Shortsightedness will destroy the effort.

An organization's vision statement articulates in clear, simple language where the organization would like to be some years in the future. It should be accompanied by a set of guiding principles for operating the organization. As a long-range target, the vision statement will help the organization align its shorter-term objectives and efforts. Any short-term objective that is not in keeping with the vision statement should be questioned. The guiding principles assist in changing the culture, because they establish the type of organizational behavior that is acceptable. Only when people abide by the organization's principles can a culture of trust and respect be established.

Both the vision statement and the guiding principles should become tip-of-the-tongue familiar to everyone in the organization. There is no reason to keep them secret, even from the outside world. Suppliers and customers should also be acquainted with them. In fact, input from all stakeholders, especially from employees, in the formulation of the vision and guiding principles is recommended.

Planning the Implementation

Implementation of Total Quality is one of the most complex endeavors an organization can undertake. It requires the cooperative efforts of many people, results in fundamental changes to the organization's culture, and requires a long-range vision for the organization. All of this suggests that careful, thoughtful planning is required, just as it is with any other complex human undertaking. It is impossible to predict the outcome of the prescribed steps when implementing Total Quality. Therefore, alternatives and recovery actions must be considered. The fundamental role of planning in the implementation process is to minimize risks, eliminate surprises, facilitate changes to the culture, and be prepared for the unexpected.

Planning by the Steering Committee begins in Step 12. This is where the implementation process is tailored to the specific organization. Will the implementation plan be based on the organization's processes, employee involvement, an organizational deployment, or some other factor? The Steering Committee must make this decision and plan accordingly.

The plan should help the organization begin implementation in small, incremental steps. It should ensure that the twenty-step implementation process is compatible with the organization and its vision. The plan should also identify the improvement projects to be undertaken by teams and provide direction to the teams. As Total Quality begins to take root, the Steering Committee will probably have to plan infrastructure changes to support team activities, possibly including changes to the organization's structure. The objective is to bring infrastructure in line with the organization's Total Quality objectives.

It should be noted that planning becomes a permanent task for the Steering Committee. It does not stop when the implementation is complete. Planning becomes a major ongoing task of the Total Quality organization's management team.

THE NEED FOR CONTINUED STUDY

Total Quality is a broad and constantly evolving subject, and there is much more to learn than is possible in the short time allocated to the Steering Committee's training session. Steering Committee members should seek out trade journals for articles that provide new, useful ideas. Steering Committee members should be committed to learning as much about Total Quality as possible. The works by and about Deming, Juran, Ishikawa, Ohno, and other authorities offer a wealth of information that will be directly applicable to the implementation and ongoing development of Total Quality. Books and articles devoted to specific elements of Total Quality, such as benchmarking and Just-in-Time manufacturing, are now widely available. Learn everything you can about the subject. Clearly, the more you know, the more effective your Total Quality initiative will be.

SUMMARY

1. The rationale for training the Steering Committee is that its members must lead the implementation, and therefore they must be equipped to do so. The training should provide a strong foundation of knowledge and credibility. It should convert holdouts on the Steering Committee and equip all members to lead the implementation.
2. The Steering Committee is typically trained by an outside consultant. It is not wise to have training provided by the leader of the Steering Committee unless this person has handled a successful implementation in another organization.
3. The training curriculum should include the following topics as a minimum: The History of Total Quality, The Total Quality Philosophy, The Rationale for Total Quality, Changing the Organization's Culture, Statistical Tools for Total Quality, Implementing Total Quality, The Role of the Organization's Vision, and Planning the Implementation.

KEY TERMS AND CONCEPTS

Continued study

Continuous improvement

Conversion of holdouts

Credibility of the leadership

Cultural change

Customer orientation

Employee involvement

Equipping the Steering Committee

Knowledge base

Management by facts

Planning

Quality first

Quality tools

Total organization involvement

Total Quality philosophy

Training curriculum

Trust

Unity of purpose

Vision statement

=== **REVIEW QUESTIONS** ===

1. Explain the rationale for training the Steering Committee.
2. Why is it usually best to bring in an outside consultant to train the Steering Committee?
3. Name four Total Quality pioneers who are important to the concept's history and describe the main contributions of each.
4. What is the rationale for Total Quality?
5. How would you respond if the top manager in your organization asked you the following question: "How can we change the culture of this organization to a Total Quality culture?"
6. List and briefly describe the purpose of the most commonly used statistical tools in a Total Quality setting.

=== **ENDNOTES** ===

1. Stephen Uselac, *Zen Leadership: The Human Side of Total Quality Team Management* (Londonville, OH: Mohican Publishing Company, 1993), 101.
2. David L. Goetsch and Stanley Davis, *Introduction to Total Quality* (Columbus, OH: Macmillan Publishing Company, 1994).
3. Frederick W. Taylor, *The Principles of Scientific Management* (New York: Harper & Brothers, 1911).
4. Kaoru Ishikawa (translated by David J. Lu), *What Is Total Quality Control? The Japanese Way* (Englewood Cliffs, NJ: Prentice-Hall, Inc., 1985), 103.
5. Ishikawa, 103.
6. M. Scott Myers, *Every Employee a Manager*, third edition (San Diego, CA: Pfeiffer & Company, 1991), 286.
7. Joseph M. Juran, *Juran on Planning for Quality* (New York: The Free Press, a Division of Macmillan, Inc., 1988), 136.
8. Goetsch and Davis.

=== **CASE STUDY 4–1** ===

Developing the Total Quality Training Curriculum at MTC

It was still early in the process, but John Lee was encouraged. The teamwork training he and the other Steering Committee members had completed was already paying dividends. Lee's managers had never worked together so well. Teamwork had been evident in the development of the Total Quality training curriculum. The consultants and the Steering Committee had worked well together in developing the curriculum. In Lee's opinion, the final version of the curriculum had just the right scope and depth.

What was especially encouraging to Lee was the progress being made by other companies that had adopted the Total Quality approach in order to maximize competitiveness. The transition to Total Quality in the United States was having a global impact.

American industry was beginning to regain market share from the Japanese in such key markets as automobiles, computers, and consumer electronics. To John Lee, this was an indication that he had set his company on the right track.

===== **CASE STUDY 4–2** =================================

Quarreling about the Training Curriculum at ESC

Fortunately for Lane Watkins, his appointment to update John Hartford on progress with the implementation had been canceled. The cancellation had given him a temporary reprieve, but it wouldn't last. Soon he would have to tell Hartford that the teamwork training for the Steering Committee had turned into a dogfight and gone nowhere. Now that his attempt to have the Steering Committee establish a curriculum had also fallen apart, he had no progress and two major failures to report.

He had begun by asking the Steering Committee members to brainstorm major topics to be included in the Total Quality training curriculum. At first there had been agreement on a few topics, such as the history of Total Quality and the Total Quality philosophy, but then someone brought up statistical tools as a topic and the brainstorming session had degenerated into an argument about statistics. Unable to get things back on track, Lane Watkins had adjourned the meeting and gone back to his office to contemplate his career and the future of ESC.

Develop the Vision Statement and Guiding Principles

The first four steps of this book have dealt with securing a firm commitment to Total Quality on the part of top management, creation of the Total Quality steering committee, building the steering committee team, and training the steering committee members in the fundamentals of Total Quality. All of this activity falls under the heading of preparation. In this chapter, the organization takes the first step beyond preparation and develops the organization's vision statement.

Many organizations, public and private, try to operate without the benefit of a long-range plan or vision. Successful people don't attempt to live their personal lives without a vision of where they want to go or what they want to become. People need direction in their lives to keep from wavering with every new circumstance. Imagine, then, a complex organization with many employees, all of whom have individual aspirations, trying to accomplish something meaningful and profitable without a guiding vision or a sense of direction. This is precisely the way many organizations, small and large, operate. This chapter shows that Total Quality organizations have an imperative to do better than this. It describes how to set organizational direction, the results that are possible when this is done, and the consequences of ignoring the issue.

DEFINITION OF ORGANIZATIONAL VISION

In the previous chapter we made the point that the vision statement summarizes the organization's long-range view. Organizations need a long-range vision because Total Quality can be fully achieved only over a long time, although there may be noticeable improvements even in the early stages of implementation. Remember that Total Quality is about fundamental changes in how things are done and how people work together. In a Total Quality organization, customers and suppliers are involved in ways never before considered, and the organization puts value on matters that have not traditionally been considered important. The implementation of Total Quality does not come together overnight. Consequently, the vision must be long enough in scope to provide a consistent course when the inevitable bumps in the road inhibit progress. Short-term attention on temporary detours can cause an organization to lose its way.

The vision statement, with its accompanying guiding principles, says, "This is where we want to be in years to come, and this is how we will conduct business in order to get there." The vision statement describes the ideal state for the organization—its values, its aspirations, and what it will ultimately be when its vision is fulfilled.

PURPOSE OF AN ORGANIZATION'S VISION

On May 25, 1961, John F. Kennedy, then President of the United States, said,

> I believe that this nation should commit itself to achieving the goal, before this decade is out, of landing a man on the moon and returning him safely to earth. No single space project in this period will be more impressive to mankind or more important for the long-range exploration of space. And none will be so difficult or expensive to accomplish.

This was a statement of Kennedy's vision for the United States, and he did a good job of developing and communicating it. Consequently, the American people understood the vision, made it their own, and ultimately the vision was realized.

In vision-driven organizations, the vision is a guiding force and a reason for being. The vision represents the beacon toward which the organization continually and consistently navigates. The accompanying guiding principles tell the organization how it is to navigate. It is analogous to a major goal for the nation, such as that set forth in President Kennedy's famous speech made a year before an American astronaut had even orbited the earth. Kennedy's visionary goal, undergirded by the guiding principles of the nation, born of the Declaration of Independence and the Constitution, was achieved eight years later, five months before the end of the decade.

It is interesting to note that by following the original American Vision, as expressed in the Declaration of Independence (Figure 5–1), and the guiding principles so well crafted by the founding fathers, as contained in the Preamble to the Constitution (Figure 5–2), the United States has fared well for over 200 years with leaders who have ranged from excellent to inept. In other words, when we know where we want to go, and when we have a set of principles to guide our actions, progress can be made even when mar-

> We hold these Truths to be self-evident, that all Men are created equal, that they are endowed by their Creator with certain unalienable Rights, that among these are Life, Liberty, and the Pursuit of Happiness. That to secure these Rights, Governments are instituted among Men, deriving their just Powers from the Consent of the Governed. That whenever any Form of Government becomes destructive of these Ends, it is the Right of the People to alter or to abolish it, and to institute new Government, laying its Foundation on such Principles, and organizing its Powers in such Forms, as to them shall seem most likely to effect their Safety and Happiness.

Figure 5–1
The Declaration of Independence: The Vision of the United States

ginal leaders are in charge. Of course, progress comes faster when good leadership is combined with a solid vision and guiding principles. So it is in any organization.

There are many examples of businesses that were started under strong leadership and that grew and prospered under the personal vision and principles of the original leader. Eventually, however, management of the business was turned over to someone new, someone with a different view of the future, a different agenda, different principles. When this occurs, the new leader tugs in one direction while the employees often pull in another. Few organizations can withstand such a situation without coming apart, and this is often what happens. This fate is not reserved for small organizations; the giant corporations are also vulnerable.

Examples of companies that have struggled with vision problems include General Motors and IBM. Founded in 1908, General Motors outperformed the competition to become the world's leading automobile producer by excelling in mass production and marketing. Alfred P. Sloan was at the helm of General Motors during its glory years (1923–1946). His successors, however, lacked Sloan's vision and his genius. Their committee-designed cars did not interest the public as much as former models had. However, because of the tremendous wealth that had been built up by the corporation, General Motors was able to plod along for years until in the 1981–1990 era, under chairman Roger Smith, events finally caught up with the automaker. When asked by *Fortune* magazine what had gone wrong, Smith responded, "I don't know. It's a mysterious thing."[1]

> We the People of the United States, in Order to form a more perfect Union, establish Justice, insure domestic Tranquility, provide for the common defence, promote the general Welfare, and secure the Blessings of Liberty to ourselves and our Posterity, do ordain and establish this Constitution for the United States of America.

Figure 5–2
The Preamble to the Constitution: The Guiding Principles of the United States

Actually, the decline of General Motors wasn't mysterious at all. The one-time world leader had lost its vision. In 1972 General Motors was ranked fourth among the world's corporations, with a stock market valuation of more than $23 billion. The three corporations that ranked higher were IBM, AT&T, and Eastman Kodak. Ten years later, General Motors had slipped one notch to fifth, although it had lost over $4 billion in valuation. By 1992, General Motors had disappeared from the top ten, sliding incredibly all the way to 40th position in the world corporate rankings. With no clearly defined and effectively articulated corporate vision, General Motors was like a rudderless ship. Consequently, its decline was predictable.

IBM was founded in 1911 and outperformed its competition in the development, sales, and servicing of business machines. Over time, the primary business machine of IBM became the computer, and the IBM vision revolved around mainframe computers.

Between 1914 and 1957, IBM was led by two Thomas J. Watsons, father and son. As IBM's leadership changed, so did the world of computers. Unfortunately, the company's vision was purely to dominate the mainframe market, even though that market was swiftly disappearing. CEO's after the Watsons failed to grasp the changes that were happening in computer technology and develop a new, more appropriate vision: personal-computer-based networks.

In 1972, IBM was the world's largest corporation, with a stock market valuation of nearly $47 billion. IBM retained this position in the 1982 rankings with a valuation of $57 billion. However, the personal computer revolution was just around the corner. By 1992, the personal computer had steamrolled IBM's mainframe vision, and the world's dominant corporation had fallen to 26th place. Even more incredibly, its valuation had plummeted to $29 billion—a drop of nearly 50%, and during a period of high inflation. Blind adherence to an outdated vision had brought an end to IBM's worldwide dominance.

In each of these cases, factors other than outdated visions were at work. Also contributing to the loss of stature of IBM and General Motors were their failure to see that the world was changing, failure to understand that customers had become more discriminating, and refusal to believe that hungry competitors that focused on customer expectations and the collective power of the workforce were a real threat. However, the outcome could have been much different. When long-term leaders who created the original vision retired, had these giants had the benefit of a forward-looking corporate vision and a set of guiding principles, the companies would not have floundered and lost their direction. Had their corporate vision been customer focused, when the competition became intense, these companies would not have been so unaware of their own myopia, but would have been alerted to the changes taking place in the world. As an example of what might have happened, a study conducted by Stanford University for the period 1926 to 1992 comparing the performance of visionary companies with that of carefully selected comparison companies revealed that the visionary companies outperformed the comparison companies by a factor of 8.6.

The comparison companies chosen by Stanford were not poor performers. In fact, they outperformed the general market by more than six to one. In other words, $1,000 invested and continually reinvested in a company in the general market in 1926 would have been worth about $415,000 in 1992. An investment of $1,000 in one of the com-

---

TOTAL QUALITY TIP

Importance of Executive Involvement

"If there is a quality council in the company, the CEO should personally sit on it and usually chair it. The CEO should also personally get into establishing quality goals, run those goals past managers and create means of measurement so that they can fulfill the quality program."[2]

Joseph M. Juran, J.D., Juran Institute, Inc.

---

parison companies over that same period would have been worth $2,671,000. Had that $1,000 been invested and reinvested in one of the visionary companies, it would have yielded $23,063,000—55 times the yield of a general market investment.[3] Clearly, vision is a major contributor to success in a competitive marketplace.

Given that an organization's vision represents a guiding force for the entire organization, providing an ideal toward which to strive, it should transcend the individual agenda of any single leader, even the CEO, thereby providing consistency through changes in leadership. The vision becomes the unifying mission of all employees, providing a reason for their work beyond the weekly paycheck.

VISIONARY LEADERS VERSUS VISIONARY ORGANIZATIONS

It is important to understand the distinction between a *visionary leader* and a *visionary organization*. We typically think of visionary leaders as people who foresaw the need for a certain product or service and set about to build or provide it. Those whose prediction of a need was correct–and to be later considered visionary it must have been–often built a great company around that product or service. Henry Ford did not invent the automobile, but he saw clearly the need to make it available to the masses. Having envisioned that need, he built the company bearing his name, which developed the means to mass produce low-cost automobiles, revolutionizing the production of automobiles forever. Ford's vision was to build a car that was easy to operate, easy to repair, and affordable by ordinary people. His Model T was that car. He produced it from 1908 until 1927, by which time other automakers had caught onto his vision. Overtaken by General Motors, Ford Motor Company became just another player in the automobile industry. As long as Henry Ford was in charge, Ford Motor Company followed his vision of building inexpensive, dependable cars. However, when Ford left, his sons succeeded him and took the company wherever their personal vision dictated. Unfortunately for Ford Motor Company, Henry Ford's sons did not have their father's vision.

Organizational vision is different from personal vision, although the distinction is subtle. An organization's vision is not about a specific product or service, but about the means of making a product or providing a service. It is a vision of the organization itself. This kind of vision is about establishing and maintaining an organization guided by principles that ensure its survival and prosperity as times, circumstances, and leaders change.

Jim Collins of the Stanford Graduate School of Business uses the following analogy to clarify the visionary leader/visionary organization concept:

> Which would be the more visionary? To be somebody who can tell you what time it is at any time, who regularly goes to the center of town and calls out "It's nine o'clock, it's nine-ten, it's nine-fifteen," and so forth. *Or* to build a clock, one that no matter what, even when I died, would always tell the time. It is independent of me. In visionary companies, people are not time-tellers; they're clock *builders*.[4]

PROPERTIES OF THE VISION STATEMENT

A well-developed vision statement must be the first item in an organization's competitiveness toolbox. A well-written vision statement with its attending guiding principles has the following properties (Figure 5-3):

1. It is easily understood by all stakeholders (employees, customers, suppliers, etc.).
2. It is briefly stated yet clear and comprehensive in meaning.
3. It is challenging yet possible to accomplish, lofty yet tangible.
4. It is capable of stirring excitement and unity of purpose among stakeholders.
5. It sets the tone for how the organization and its employees conduct their business.
6. It is not concerned with numbers.

The vision statement must be crafted in such a way that all employees can relate to it, and, in so doing, execute their work in a manner and direction that is consistent with its meaning and objectives.

DEVELOPING THE ORGANIZATION'S VISION AND GUIDING PRINCIPLES

While it is *possible* for the leader of an organization to single-handedly develop the vision and guiding principles for an organization, this is not the best approach. There are three important reasons for this contention. First, the CEO's unilaterally developing vision sends a strong message to senior managers—the Steering Committee in this case—that their input is not wanted or valued. Second, it suggests—wrongly—that the leader acting alone can develop a better vision statement than the Steering Committee members working collectively. Finally, it is easier for Steering Committee members to buy into the organization's vision when they have had a hand in its creation. Hence, the best approach is to make development of the organization's vision and guiding principles a project of the Steering Committee.

Developing the vision and guiding principles is best done away from the normal workplace in casual surroundings similar to those used for the teambuilding and training sessions so that discussions of where the organization should go and the principles to be observed in getting there are not interrupted by workaday concerns. Concentration

Figure 5–3
Properties of an Effective
Vision Statement

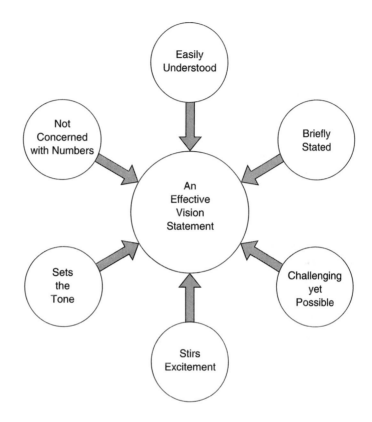

can be difficult when meetings take place in close proximity to the offices and telephones of participants.

It is appropriate for the organization's leader to facilitate the development of the vision statement and guiding principles. With the teambuilding and training sessions behind them, the Steering Committee members should be well prepared to work together in developing the vision and guiding principles. This will be the Steering Committee's next big test. As before, if the leader is not comfortable serving as facilitator, an outsider should be brought in to serve that function.

Although the leader should not develop the vision alone, it is appropriate for the leader to come to the meeting prepared to present a "trial balloon" to get discussion started. The leader's draft ideas can also serve to stimulate dialogue if the session is handled correctly. However, there is the danger that the participants might take the leader's draft as *the* vision statement, requiring only their acceptance. Under no circumstances should this be allowed to happen. The leader must make it clear that the draft simply represents his or her thoughts, that it is by no means the final vision but is provided to initiate dialogue.

The leader should draw ideas from the Steering Committee and must never display a negative or defensive attitude. Leaders who cannot do this should bring in a facilitator. The point here is that everyone's ideas, thoughts, and arguments are wanted because this kind of free and open discussion will ultimately bring out the best input. It is also through this kind of discussion, with all points of view being freely contributed, that understanding and ownership of the ultimate vision statement will develop.

The leader must come to the meeting prepared, as should the rest of the Steering Committee members. It is essential that the leader be informed concerning what the vision statement and guiding principles are all about. The leader, or facilitator, should begin by instructing the Steering Committee on the structure of the vision statement.

The vision statement consists of two basic elements. The first is the vision statement itself, which is a description of the ideal state for the organization. This serves as the target toward which all employees will work. It should energize all employees, giving them a high-minded reason for their work and providing a well-directed focus for all their activities. The following are examples of acceptable vision statements:

- We will be recognized by our customers as the highest quality producer of _____ in the world.
- We will be seen by our customers as the supplier of choice for all our products.

The following statements are examples of unacceptable descriptions:

- Our reason for being is to make money for our stockholders.
- Our ultimate objective is to maximize profits.
- We will build a revolutionary new _____.

What is wrong with these statements? Commercial organizations must make a profit, and making money is certainly an honorable objective. But simply making a profit cannot be the ideal state of the organization. Such a statement would tend to encourage behavior that is penny wise but dollar foolish. Managers whose performance is graded against the bottom line tend to apply short-term and shortsighted strategies. They tend to cut corners, push employees beyond their limits, eliminate training, and neglect equipment maintenance.

The vision statement should not focus on a specific product, as the third example does. Such visions are too shortsighted. When this statement is used, what happens once the vision is achieved? When the revolutionary product is put into production, what then? A product-focused vision statement is too confining, too narrow, and too short-term in its orientation. In addition, it causes the organization to put all its eggs into one basket.

Notice that the examples of acceptable vision statements focus on the organization's customers. The first statement says that the organization will prove to customers that its product is the best available. The second statement says that the organization's products will be the customer's choice from among all competitors. This organization might be a manufacturer, a distributor, or even a retailer. In any case, the vision is to be the provider of choice for customers who have a given set of needs and preferences.

Taken by themselves, these statements raise some questions. How will the organization get the customers to view it as the supplier of choice? How does the organization produce the highest quality products? Such questions are answered by the second ele-

---

TOTAL QUALITY TIP

Why Total Quality Sometimes Fails

*"When TQM programs fail, it is because they are mounted as programs, uncon-
nected to business strategy, rigidly and narrowly applied, and expected to bring
about miraculous transformations in the short term without top management lifting much of a finger."[5]*

Rosabeth Moss Kanter, Harvard Business School

---

ment of the vision statement: the guiding principles. These principles will guide the
actions of employees as the organization works to accomplish its vision. If the vision
statement is customer oriented, the guiding principles must likewise relate to customers. Examples of such principles are as follows:

■ All of our activities will be conducted in accordance with the highest ethical standards.

■ Quality will always be our first and primary concern.

■ We will always be conscious of the fact that everything we do affects our customers.

There should also be principles that guide the organization's relationship with its
employees, since without satisfied employees there will not be satisfied customers.
Examples of such principles are as follows:

■ We will always treat our employees with respect, honesty, and fairness.

■ We see our employees as a major source of intelligence, welcome their input and
make use of their ideas.

■ It is the function of management to give employees the support they need to succeed by providing the appropriate organization, facilities, resources, and training.

In North America, customer satisfaction involves more than meeting a customer's
expectations. There are also expectations concerning the social responsibilities of organizations. Consequently, it is appropriate to spell out the social responsibilities of the organization as part of the guiding principles. Examples of such principles are as follows:

■ We will always be a good corporate citizen in our community.

■ We will never knowingly cause harm to the environment, our neighbors, or our
employees.

It is not necessarily recommended that you adopt any or all of these examples in
your vision statement. They are provided as examples. Your set of guiding principles
should reflect the specific vision and values of your organization. Figure 5–4 is an example of a complete vision statement.

With the vision statement and its guiding principles in place, the organization will have a target and a set of rules to operate by as it progresses toward the target. Once the vision statement and its accompanying guiding principles have been developed, the organization can take the next critical steps. These steps involve developing objectives and strategies for accomplishing the objectives (Step 6).

ALL ORGANIZATIONAL ACTIVITY MUST SUPPORT THE VISION

It is important to understand that once the vision statement is established, every action taken by the organization should support it. It does not make sense to have one part of the organization going off in one direction while another part goes in the opposite direction. Nor does it make sense to continue any function or activity that does not support the vision. A question employees should frequently ask themselves is, "Are my actions contributing to progress toward accomplishing the vision?" Whenever the answer to this question is no, a way must be found to eliminate the action or bring it around to the proper heading. Such refinements fall under the heading of continuous improvement, a basic principle of Total Quality.

The Institute for Corporate Competitiveness (ICC) will be recognized by customers as the supplier of choice for management training products that are the best in the world. In seeking to realize this vision, ICC will be guided by the following principles:

- All activities will be conducted in accordance with the highest ethical standards.
- High quality and reasonable prices will be primary concerns.
- Quality will always be defined by our customers.
- Customers, employees, and suppliers will be treated with respect, honesty, and fairness.
- All stakeholders will be empowered to help improve our performance continuously.
- Employees will be given the support and resources necessary to perform their jobs effectively.
- Continuous improvement will be the norm.
- Management will provide employees a safe and healthy workplace.
- We will be good corporate citizens in the local, state, and national communities.

Figure 5–4
Sample Vision Statement and Guiding Principles

In order for all activities to support the vision, everyone in the organization must know what the vision and guiding principles are, what the supporting objectives are (discussed in the next chapter), and how his or her work relates to them. For this reason, successful organizations go to great lengths to acquaint all employees with the vision, principles, and objectives and reinforce their knowledge at every opportunity. The vision, principles, and objectives should be printed and distributed to all employees. The leader and the Steering Committee members should talk to employees about the vision to make sure it is thoroughly understood and accepted. As progress is made, it should be reported to employees, talked about, and displayed prominently. In an organization that has implemented Total Quality, employees at all levels should be able to paraphrase the organization's vision. Without a thorough understanding of the vision, employees have no way of knowing if their work is supportive, irrelevant, or counterproductive. This is the case when an organization cannot communicate too much. In addition to communicating the vision to employees, the organization should communicate it to customers, suppliers, the community, the industry, and the world.

SUMMARY

1. An organization's vision is a long-range view of where it wants to go and the rules it will follow in getting there. The vision statement describes the ideal state for the organization; its values, aspirations, and what it will ultimately be when the vision is fulfilled.
2. The vision serves as a beacon toward which the organization is always moving. The guiding principles that accompany the vision tell all employees in the organization how they should navigate the course.
3. Visionary leaders and visionary organizations are not the same. Visionary leaders in business are typically people who foresaw the need for a product or service and built a business around it. Visionary organizations build their visions around the means to make a product or provide a service. An organization's vision is about establishing and maintaining an organization guided by principles that ensure its survival and prosperity.
4. An organization's vision should have the following qualities: easily understood, briefly stated yet clear and comprehensive, challenging yet possible to accomplish, capable of stirring excitement and unity of purpose, tone-setting, and not concerned with numbers.
5. An organization's vision and guiding principles should be developed by its Steering Committee. Developmental meetings may be conducted by either the organization's leader or an outside facilitator. Once the vision and guiding principles have been developed, all of the organization's activities should support them.
6. Everyone in the organization must know what the vision and guiding principles are, and all organizational activity must support the vision.

KEY TERMS AND CONCEPTS

Briefly stated

Challenging yet possible

Easily understood

Guiding principles

Ideal state

Lofty yet tangible

Not concerned with numbers

Organizational vision

Sets the tone

Stakeholders

Steering committee

Stirs excitement

Support the vision

Unity of purpose

Vision

Visionary leaders

Visionary organizations

REVIEW QUESTIONS

1. Define the term *organizational vision*.
2. Explain the purpose of an organization's vision.
3. Explain the distinction between *visionary leaders* and *visionary organizations*.
4. List the essential properties of an organization's vision.
5. Describe the process for developing the vision statement and guiding principles.
6. What is meant by the statement, "All organizational activity must support the vision"?

ENDNOTES

1. Carol J. Loomis, "Dinosaurs?" *Fortune,* May 3, 1993, 36–42.
2. Catherine Romano, "Report Card on TQM" (interview), *Management Review,* January 1994, Vol. 83, No. 1, 23.
3. Loomis, 36–42.
4. Tom Brown, "On the Edge with Jim Collins" (interview), *Industry Week,* October 5, 1992, 12–20.
4. Romano, 23.
5. Romano, 23.

CASE STUDY 5–1

Developing the Vision at MTC

John Lee had thought long and hard before deciding to facilitate the session in which MTC's Steering Committee would develop a vision statement and a set of guiding principles. He had discussed the matter with MTC's implementation consultant and with all Steering Committee members. The consensus was that Lee could and should facilitate the vision-development session.

Lee prepared thoroughly for the session. He met with MTC's implementation consultant to discuss logistical concerns and developed a list of ideas to be used to initiate discussion. When the day planned for the session arrived, Lee was ready.

Discussion began in earnest. Several Steering Committee members had ideas about what MTC's vision should be. One member wanted a vision that focused on a specific product. Another member wanted to change the company's direction and drop government markets from MTC's vision. Lee let anyone with an idea express it and encouraged open discussion of all ideas.

When the discussion began to lag, Lee distributed a handout entitled "Characteristics of a Good Vision Statement." MTC's implementation consultant had helped Lee develop the handout. The desired characteristics included the following: easily understood, briefly stated, challenging yet possible, not concerned with numbers, tone-setting, and able to stir excitement.

With these characteristics to guide them, the participants settled into a focused and productive discussion. Lee had prepared a rough draft of a vision statement to use as a discussion starter, but soon found it wasn't needed. The group decided that MTC should become a world-class supplier of dual-use products (products that have both military and civilian use, such as aircraft power supplies).

Once the vision statement had been completed, the group began the task of developing the accompanying guiding principles. Lee led the group in a brainstorming session to identify the characteristics associated with world-class companies. Terms suggested included teamwork, employee involvement and empowerment, quality, customer focus, creativity, scientific measurement, and many others. These characteristics were then transformed into MTC's guiding principles.

By the end of the day, the Steering Committee had developed a powerful vision statement that included a comprehensive set of guiding principles. Later that night, Lee sat in his den contemplating the day's events. He felt excited and invigorated. Now that MTC knew where it was going and what it wanted to be, he was anxious to get started.

═══════════ **CASE STUDY 5–2** ═══════════════════════════════

Vision Problems at ESC

Development of a vision was not going well at ESC. The CEO, John Hartford, had turned down Lane Watkins' request to rent an off-site location for a vision development session. Consequently, ESC's Steering Committee members had been continually interrupted by secretaries with messages that couldn't wait and subordinates with problems that demanded immediate attention. But constant interruptions were only part of the problem. Steering Committee members had questioned the entire concept. The group consensus was expressed by one member who asked, "How can we develop a vision statement for ESC? We're just department heads. There is not even one executive-level manager among us."

Watkins had countered by proposing that they develop a draft document and forward it to their top managers for approval. This had gotten the meeting started, but not much good had come of it. In fact, as Watkins reflected on it now, he realized that the meeting had probably done more harm than good. Rather than culminating in a clear vision for ESC, it had ended in disarray. The only thing that was made clear was that ESC is a company without direction, floundering and confused.

Watkins had tried to bring some structure to the meeting by distributing a handout that summarized the characteristics of a good vision statement. It had not helped much. One member suggested that ESC adopt the following vision statement: "At ESC we don't know who we are, where we are going, or how we plan to get there. We don't trust employees or suppliers, and our customers don't trust us."

After this suggestion, Watkins had lost control and angrily dismissed the meeting. Now, looking back on the meeting, he realized that the proposed vision statement, although suggested in jest, had been right on target. He resolved to begin sending out his resume the next day.

Set Broad Objectives

===== MAJOR TOPICS =====

- Rationale for the Employment of Broad Objectives
- The Nature of Broad Objectives
- The Difference between Management by Objectives and Vision with Broad Objectives
- Cautions Concerning Broad Objectives
- Establishing the Broad Objectives
- Deployment of Broad Objectives

In Step 5 the Steering Committee created the organizational vision and established the guiding principles by which the organization would operate. Step 6 takes the Steering Committee through the next logical activity in the planning process: setting broad objectives that establish the strategy the organization will use to achieve its vision. In many organizations this is called *strategic planning*. It is a necessary step when implementing Total Quality, even if the organization has never employed strategic planning.

We have chosen to use the term *broad objectives*, but the reader should understand that the term *goal* is interchangeable and could be used in this context. Similarly, when others speak of *macro goals* or *macro objectives*, the meaning is typically the same as that of *broad objectives*.

RATIONALE FOR THE EMPLOYMENT OF BROAD OBJECTIVES

In a vision-driven organization, setting broad objectives is the next step after establishing the vision and guiding principles. The vision identifies the end and the broad objectives define the means to that end. Broad objectives articulate the strategy that will be employed to achieve the vision. By establishing the broad objectives necessary to achieve the vision, a road map is developed that will help the Steering Committee establish pro-

---

TOTAL QUALITY TIP

Goals as Targets

A goal is an aimed-at target, an achievement toward which effort is expended.[1]

Joseph M. Juran

---

jects for improvement teams—projects that support the vision. The broad objectives also provide a sense of direction for all employees, thereby minimizing wasted effort on tasks that bring the organization no closer to achieving its vision.

All organizations need a vision, guiding principles, and a set of broad objectives. The alternative would be to let each department decide what it should work on. We have already seen that individual departments are prone to develop priorities that benefit the department, sometimes even to the detriment of the total organization. This cannot be allowed in the Total Quality organization.

THE NATURE OF BROAD OBJECTIVES

In the last chapter, it was explained that the vision statement represents a long-range view, an ultimate target. The vision establishes where the organization hopes to be or what it hopes to become at a point in the future. There is no suggestion of *how* the vision is to be achieved. This is where the broad objectives come into play. These objectives represent the best thinking of Steering Committee members concerning what must be accomplished in order to arrive at the ideal state represented by the vision. It can be said then, that the vision statement establishes the ultimate target, and the broad objectives provide the strategy for achieving the ultimate target.

Consider the example of Tropic Breeze Company, a manufacturer of ceiling fans. Determined competition from the Pacific Rim Company has eroded the market share historically enjoyed by Tropic Breeze. The CEO of Tropic Breeze has decided to implement Total Quality to help the company regain and increase market share. Tropic Breeze's Steering Committee created the following vision statement: "Tropic Breeze will become the international supplier of choice for ceiling fans."

Steering Committee members agreed that this statement represents the ideal state for Tropic Breeze, but initially they did not think through what would have to be done in order to achieve this vision. The broad objectives developed were as follows:

1. The aim of Tropic Breeze is *customer delight* with all of our products.
2. We will produce the most satisfying and reliable fans on the market at prices equal to or below the competition's.
3. We will aggressively seek input and feedback from our customers and suppliers.

These three broad objectives form the underlying strategy that Tropic Breeze plans to use to become the "international supplier of choice" for its products. Tropic Breeze's leadership recognized that to be the supplier of choice, customers would have to be not just satisfied, but delighted with their fans. For this to happen, their fans had to be better than the competition's fans in all features important to customers, yet priced at or below competing prices. Tropic Breeze's strategy, then, was to produce the best fans at competitive prices. There are some important factors to note in Tropic Breeze's broad objectives:

- The objectives describe the organization's strategy.
- The objectives flow from the vision and support it in all respects.
- The objectives apply to the entire organization rather than one or two departments. Everyone, regardless of function within Tropic Breeze, can relate to and have an impact on these objectives.
- The objectives are nonrestrictive. That is, there is nothing about them that would keep any individual or any group from doing its absolute best.
- Although the objectives form the strategy, they do not reveal specific details of *how* they will be accomplished. Instead, the objectives form a bridge between the vision and the specific tactics that will be developed at the department and team levels. They lead to the development of the projects (tactics) that will be deployed throughout the company.

As shown in Figure 6–1, development of the objectives is the second of three steps. The first is creating the vision. The third is translating the objectives into activities throughout the entire organization (this activity is covered in Step 13).

If these goals seem somewhat fuzzy—and they might to people who are accustomed to more rigid, detailed objectives—consider a company that set just one broad objective to support its vision. The company is Motorola, and its objective said only that, no matter what an employee was doing, he or she should "try to get better by a factor of ten in the next five years." Motorola is now about 100 times better than when the objective was first stated. Although the Tropic Breeze example used no numbers, there is certainly nothing wrong with inserting numbers into broad objectives as long as they do not prescribe limits.

Motorola envisioned a five-year period for substantial (tenfold) improvement in all its functions. Motorola did not see the five years as an absolute mandate, however, and was happy to make the desired improvements in just three. On the other hand, had Motorola's objective been set at a factor of two rather than ten, it might have turned out to be limiting. When performance doubled, there could have been a relaxation of motivation to continue improvements. Broad objectives stated in terms of reduction by X percent or improvement by X times are acceptable, but be sure that optimal limits are set. When in doubt, leave the numbers out of broad objectives and put them in the specific tactics developed later.

It is essential that the Total Quality organization address quality in its broad objectives. Consideration of customer satisfaction, the cost of poor quality, improving a prod-

Figure 6–1
Steps in Developing and
Carrying Out the Organi-
zation's Vision

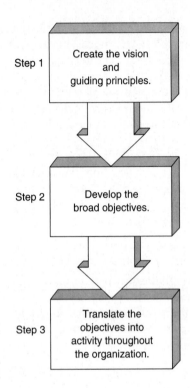

Step 1 — Create the vision and guiding principles.

Step 2 — Develop the broad objectives.

Step 3 — Translate the objectives into activity throughout the organization.

uct or service, and eliminating waste are important for every organization. Any set of broad objectives that does not address quality is incomplete and ill-conceived. Typically, the objectives of traditional companies stress money—orders, sales, costs, profit, bottom line, and so on. We are just beginning to understand that attempting to correct financial problems without first correcting their underlying causes is futile. Quality objectives can have a greater impact on financial results than pure financial objectives, because quality-oriented objectives target the root causes of financial woes. Many organizations have yet to realize this.

THE DIFFERENCE BETWEEN MANAGEMENT BY OBJECTIVES AND VISION WITH BROAD OBJECTIVES

The Total Quality organization has a vision that is supported by a set of broad objectives, which in turn is deployed throughout the organization in the form of projects, tasks, and tactics for employees and teams of employees. How does this differ from Management by Objectives (MBO)? When Peter Drucker invented MBO in 1954, he did not intend it to be what it has become. Drucker's idea was that superior results could be obtained by allowing people to work toward mutually agreed-upon objectives while controlling their own actions. MBO was intended to be antibureaucratic in nature. However, over time, MBO became weighted down with bureaucratic control. Rather than fostering self-control of work, it became a means of tightening management controls on those doing the work. It is this manifestation of MBO that Dr. Deming argued against, and so do we.

MBO, as it is typically practiced, is very different from the vision-driven Total Quality approach. The differences in these two concepts are illustrated in Figure 6–2 and discussed in the following paragraphs.

In the Total Quality organization, senior management establishes a vision that clearly articulates what the organization wants to be. There is no corollary in the MBO organization.

In the vision-driven organization, senior management translates the vision into a few key strategic, customer-oriented objectives that must occur in order to realize the vision. In the typical MBO organization, the key strategic objectives are oriented toward fiscal issues such as the bottom line, and they are typically shorter in scope of time. MBO objectives tend to be numerous and very specific.

In the vision-driven organization, the broad objectives are translated into specific tactics, sometimes by the individuals who do the work, but increasingly by the Steering Committee with the involvement of individuals who do the work. The details of *how* the projects are to be carried out are typically left to the employees, because this is their area of expertise. Such a process ensures that all the projects are aligned with the broad objectives and the vision. With MBO, all managers at all levels are required to have a list of objectives to be accomplished during the next year. The usual procedure is for the manager to develop a list of six to ten objectives that he or she thinks are important, then negotiate their acceptance with his or her supervisor. Negotiations might relate to schedules, numeric results, or even the deletion or addition of objectives. Once negotiated, a manager's objectives presumably give direction to his or her activities, as well as those of subordinates. The intent is that objectives be established at each level to support those of the next higher level. What actually happens is quite different. With MBO, the

Figure 6–2
Differences between
Management by Objectives
and Vision with Broad
Objectives

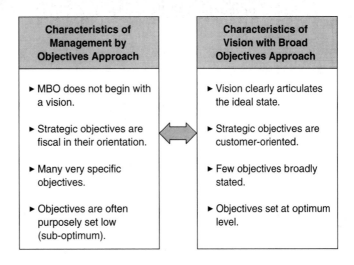

Characteristics of Management by Objectives Approach	Characteristics of Vision with Broad Objectives Approach
▶ MBO does not begin with a vision.	▶ Vision clearly articulates the ideal state.
▶ Strategic objectives are fiscal in their orientation.	▶ Strategic objectives are customer-oriented.
▶ Many very specific objectives.	▶ Few objectives broadly stated.
▶ Objectives are often purposely set low (sub-optimum).	▶ Objectives set at optimum level.

myriad objectives frequently wind up pulling the organization in several directions at once. This is especially true between departments, as was discussed in earlier chapters.

The specific objectives promulgated under MBO differ from the broad objectives of Total Quality in several key ways. The most important ways are summarized as follows:

■ MBO objectives are often department-oriented and therefore advance a departmental agenda rather than that of the overall organization. The Total Quality organization's vision-guided objectives have the single purpose of supporting the vision. As a result, everyone pulls in the same direction.

■ MBO objectives are typically concerned with individuals and are directly linked to performance against specific objectives. Performance appraisals in a Total Quality organization focus more on team performance.

■ MBO objectives tend to be short term, focusing on something that can be achieved in less than a year. Total Quality objectives have a longer frame of reference, and hence provide continuity of purpose and effort.

■ MBO objectives tend to be suboptimal because responsible personnel often intentionally set easily achieved objectives so as to look good on performance evaluations. The Total Quality organization's vision-driven objectives are liberating. They don't put limits on success.

CAUTIONS CONCERNING BROAD OBJECTIVES

Following are several cautions that should be understood concerning broad objectives and how they are employed. The cautions are summarized in Figure 6–3.

■ Broad objectives should be few in number, typically three to five.

■ The language should be kept simple so that it will be readily understood by *all* employees.

■ Every objective should be something that must be achieved if the vision is to be realized.

■ If numbers are part of the objective, they must not limit expectations and eventual results.

■ Developing broad objectives is strategic planning, and as such is a means to an end, not the end itself. Do not become bogged down in endless analysis and continual meetings to establish the objectives.

■ Broad objectives are not used in the employee-appraisal process.

■ Broad objectives must be crafted in such a way that they relate to all employees, never to a single group or department.

■ Broad objectives should require some stretch on the part of employee teams and management, but should not be made impossible. If objectives are perceived as being unachievable, they will be ignored.

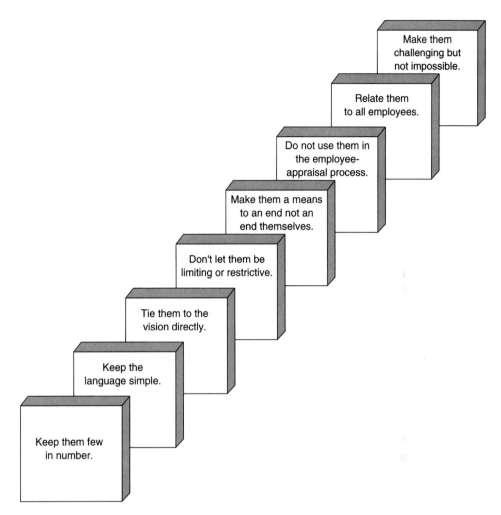

Figure 6–3
Cautions to Observe when Developing Broad Objectives

ESTABLISHING THE BROAD OBJECTIVES

By this point in the implementation process, Steering Committee members should be comfortable working together as a team. This is the mode in which the broad objectives should be developed. It is important that the team devote the time necessary to the formulation of these objectives, and that it do so in an atmosphere that is conducive to the free interchange of ideas. This typically means the Steering Committee should meet away from the office for the one or two days it will take to develop the broad objectives.

TOTAL QUALITY TIP

Objectives Must Be Achievable

In his book, Thriving on Chaos, *Tom Peters uses the analogy of teaching a person to swim: If put into a pool of water just slightly over his head, a person can bob up and down off the bottom to get air, and will probably learn to swim. But put the same person into a much deeper pool and he will probably drown.*[2]

The CEO should lead the meeting. The use of a facilitator is optional. A facilitator is probably a good idea, particularly if it is the first time the team has developed broad objectives. The facilitator can help the team get past just recreating the same kinds of objectives they have used in the past. The leader and all of the Steering Committee members should come to the meeting well versed in the intent of the meeting and the nature and function of broad objectives. Each participant should have devoted some serious thought to the vision and what must happen in order to achieve it. We are not suggesting that anyone, including the leader, come to the meeting with preconceived notions of what the broad objectives should be. Rather, all participants should be well informed and prepared to become active participants in the process.

The vision statement should be prominently displayed in the meeting room for easy reference. The first order of business is a review of the planned agenda. This will give the team a schedule and help keep participants focused. The next task is a review of the vision and the thinking behind it. This is followed by a review of the process of developing broad objectives that support the vision and deploying those objectives. This review will set the stage concerning the purpose of the meeting. With these preliminary steps out of the way, the team is ready to begin developing objectives.

The process of developing the broad objectives can take several forms. Which approach is used should be determined by the team. One method is to begin by brainstorming potential objectives. Proposed objectives are written on flipcharts and the filled pages are taped to the wall for easy viewing. The leader or the facilitator should ensure that everyone contributes and that there is no discussion of the ideas that are presented—at least not yet. When the flow of ideas runs its course, the team consolidates similar objectives, culls out those that obviously don't belong, and considers the rest. Eventually consensus will form around several objectives. The Steering Committee then refines these final objectives as needed to ensure that they support the vision.

Another approach is to go around the table, taking one potential objective from each participant and repeating the process until there are no additional objectives proposed. This works best when the team members have been asked before the meeting to come prepared with a list of proposed objectives. After the proposed objectives are all collected and displayed on the wall, the consolidation/culling/selection steps are applied. The final objectives are discussed and refined as before.

A third method is to engage the team in a discussion of what must be done in order to achieve the ideal state set forth in the vision. Suggestions are recorded. The resulting activities become the broad objectives.

Dr. Juran lists five main activities involved in the process of setting goals (Figure 6–4).[3] Although written for the process of optimizing product design, the five main activities apply to any form of goal-setting by groups.

- Assemble the *Inputs* (by one of the three methods just explained).
- Find the *Optimum* (by consolidating, culling, selecting).
- Resolve *Differences* (by discussion and refinement).
- *Decide* on the Goal.
- *Publish* the Goal.

We have already discussed the nature of broad objectives and the precautions that should be observed in developing them. It is important to keep these in mind, because it is easy to slip back into the familiar pattern of developing typical departmental objectives or concentrating on fiscal or productivity issues. Be certain that the objectives selected are broad in scope so that they apply to the entire organization, and are in no way limiting or suboptimal.

Remember that the function of broad objectives is to help the Steering Committee establish projects and goals for subordinate teams and employees, and to help the total organization. It makes no sense to establish objectives and then treat them as secret information, but this is exactly what many organizations do. Such organizations may be

Figure 6–4
Steps in Developing
Objectives

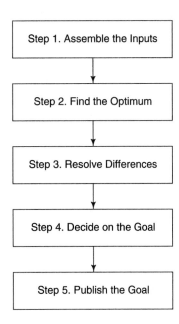

trying to keep their objectives out of the hands of competitors. In reality, competitors can predict these things anyway. In fact, keeping organizational objectives hidden from employees who will have to accomplish them is a sure way to help the competition. Employees who are kept in the dark about what the organization needs to do cannot help the organization do it effectively, and may even be working counterproductively.

Total Quality organizations use every means at their disposal to publicize their vision, broad objectives, and corresponding project and team goals. They use employee meetings, company newsletters, bulletin boards, one-on-one conversations, and numerous other strategies for communicating the organization's vision and objectives to all stakeholders. In addition, Total Quality organizations make certain that the employees know about progress and problems along the way. There should be no secrets concerning the vision and objectives.

DEPLOYMENT OF BROAD OBJECTIVES

The final step in developing the broad objectives is deployment. While not strictly part of the objective-setting process, deploying objectives is a critical step. The Steering Committee will use the broad objectives in establishing projects for subordinate teams throughout the organization. The Tropic Breeze example used earlier illustrates this point. The vision statement and broad objectives for Tropic Breeze are as follows:

Vision Statement:
Tropic Breeze will become the international supplier of choice for ceiling fans.

Broad Objectives:
1.   The aim of Tropic Breeze is customer delight in all of our products.
2.   We will produce the most satisfying and reliable fans on the market at prices equal to or below the competition's.
3.   We will aggressively seek input and feedback from our customers and suppliers.

Tropic Breeze's broad objectives state what must be done in order to realize the vision, but do not indicate how the objectives are to be accomplished. This step is deployment, or the translation of the objectives into specific projects for teams.

Tropic Breeze's management team understands that in order to satisfy its first objective, customer input must be collected. To collect customer input, the Steering Committee established a cross-functional team. The team consists of an engineer, a manufacturing assembler, a member of the customer service department, a sales representative, and a member from the warranty department. Team members were updated on the company's vision and broad objectives. They were then given the project of collecting customer input, comments, suggestions, and complaints. The team was asked to use a four-month schedule and report progress to the Steering Committee monthly. The team was asked to begin by developing a plan for accomplishing the task. It was given a target of one month to complete the plan. The Steering Committee volunteered its services, and the services of anyone else in the company needed to assist with the plan.

After reporting progress to the Steering Committee to make sure it was on the right track, the team presented its plan. The Steering Committee approved the plan, and the team was cleared to begin the project. After three months the team had collected and summarized customer input. It found that customers were generally satisfied with Tropic Breeze's products, but not delighted. Customers thought the competition offered fans with a better value/price ratio. There were complaints about the lubrication requirement and how difficult it was to accomplish. Customers wanted more colors than the standard Tropic-Breeze Brown. Wobble was mentioned frequently as an annoyance. Many customers thought competing fans were quieter. Warranty repairs were infrequent, but they were time consuming when needed. Complaints about customer service focused primarily on the difficulty in getting through on Tropic Breeze's 800 line, and then having to wait for a technician.

With this information, the Steering Committee was able to set up several improvement teams to attack the problems identified. An engineering team was assigned the project of eliminating wobble. Another team was assigned the project of eliminating motor and blade noise. Another was to consider options for lubrication. Other teams were deployed for the remaining issues, including those related to customer service and warranty repair. While all of this was going on, a team which included representatives from key suppliers, engineering, manufacturing, customer service, and finance was charged with the task of making the fans easier to assemble, while at the same time reducing scrap, the anticipated output being higher quality and lower cost. Just three broad objectives had been translated into a host of team projects, and it was clear that the possibilities for additional projects were unlimited.

Following is a list of considerations to be understood when attempting to deploy objectives.

- The broad objectives create a bridge between the vision and the teams that will undertake actual projects in support of the vision, as illustrated in Figure 6–5.

- Deployment does not mean exhortation by management to subordinates. It is accomplished through specific projects undertaken by teams.

Figure 6–5
Broad Objectives as a Bridge between the Organization's Vision and Team Projects

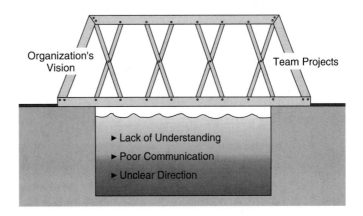

Organization's Vision

Team Projects

► Lack of Understanding

► Poor Communication

► Unclear Direction

- In the deployment process, the Steering Committee translates the broad objectives into specific projects or goals for teams or individuals.
- The Steering Committee assigns the specific projects to the teams. The teams develop plans for accomplishing their assignments.
- Deployment projects represent tactics for accomplishing the strategy (the broad objectives) which, in turn, lead to securing the vision.

Deployment is addressed in greater detail in Step 13.

SUMMARY

1. The rationale for the employment of broad objectives is that they identify ends. They translate the vision into more specific details for realizing the vision without getting into how-to issues.
2. The nature of broad objectives is such that they represent the best thinking of the Steering Committee concerning what must be accomplished in order to arrive at the ideal state represented by the vision. Broad objectives provide strategic insights for achieving the ultimate target.
3. There is a subtle but important difference between Management by Objectives (MBO) and the vision with broad objectives approach. MBO does not begin with an organizational vision. In addition, MBO is a narrowly focused concept that allows individual departments to function in ways that may not serve the overall organization. The vision with broad objectives approach ensures that all employees, all teams, and all departments are pulling in the same direction.
4. Broad objectives should be few in number, written in simple language, and crafted in such a way that they apply to all employees.
5. The organization's broad objectives are developed by the Steering Committee. The CEO should run the meeting in which the objectives are developed; the use of a facilitator is optional. Brainstorming and other related approaches are used to collect input from Steering Committee members.
6. The process of developing objectives involves five steps: assemble the inputs (by brainstorming and recording); find the optimum; resolve differences; decide on the goal; and publish the goal.
7. Once the broad objectives are developed they must be deployed. This is done by assigning specific improvement projects to teams, having the teams develop plans for making the improvements, implementing the plans, and monitoring progress.

KEY TERMS AND CONCEPTS

Assemble the inputs	Decide on the goal
Broad objectives	Deployment of objectives
Consolidating	Discussion
Culling	Final objectives

Find the optimum   Refinement
Macro goals   Resolve the differences
Macro objectives   Selecting
Management by Objectives   Strategic planning
Nonrestrictive objectives   Vision with broad objectives approach
Publish the goal

REVIEW QUESTIONS

1. What is the rationale for the employment of broad objectives?
2. Describe the nature of objectives in a Total Quality organization.
3. What is the difference between the MBO approach and the vision with broad objectives approach?
4. List cautions to be observed when developing broad objectives.
5. Describe the process used by the Steering Committee to develop broad objectives.
6. List the steps in the process of developing objectives.

ENDNOTES

1. Joseph M. Juran, *Juran on Planning for Quality* (New York: The Free Press, a Division of Macmillan, Inc., 1988), 137.
2. Tom Peters, *Thriving on Chaos* (New York: Harper-Collins Publishers, 1991), 601.
3. Juran, 153.

CASE STUDY 6–1

Setting Broad Objectives at MTC

John Lee pointed to MTC's new vision statement that was taped to the wall. It read, "MTC will be a world-class supplier of dual-use technologies." The Steering Committee had convened for the purpose of setting the broad objectives that will translate MTC's vision into action at the team level. Lee had prepared the Steering Committee members well. Each member understood the rationale for employing broad objectives and the cautions to observe in developing them.

The brainstorming of potential objectives began in earnest, with John Lee recording all ideas on large flipcharts and taping the pages to the wall. Ideas flowed freely. When the brainstorming phase of the meeting had run its course, Lee began discussions to cull out unacceptable objectives and to find the optimum objectives. Several ideas were consolidated. This narrowed the list to five. One objective was dropped altogether because it was too narrow. The remaining four objectives were discussed at length and refined several times.

By the end of the day, MTC's Steering Committee had three clearly stated, well-refined, broad objectives. MTC's objectives read as follows:

1. MTC will achieve complete customer satisfaction in all of its product lines.
2. MTC will produce high-quality products at prices equal to or below the competition's.
3. MTC will aggressively seek input and feedback from customers and suppliers.

CASE STUDY 6-2

What Are the Objectives at ESC?

Lane Watkins knew he couldn't ask ESC's Steering Committee to develop broad objectives without a vision statement on which to base them, but where was he going to get one? To solve his dilemma, Watkins decided to take the pragmatic route and develop a vision statement himself. After attempting unsuccessfully to get an appointment with his CEO, John Hartford, for over a week, Watkins decided to press on without executive approval. He knew from his study of Total Quality how a well-written vision statement should read. After several false starts he finally settled on a vision statement that read:

"ESC will be the supplier of choice for engineering services statewide."

Watkins knew he had developed a vision that is limited in scope. He didn't envision ESC as an international, national, or even regional supplier of choice. In fact, he had his doubts that ESC could achieve the vision, even in the limited market of one state. Watkins knew that if he broadened the scope of the vision the Steering Committee would laugh him out of the room.

Armed with a vision statement he could defend—although it had been developed without the involvement or approval of ESC's executive-level managers—Watkins called a meeting of the Steering Committee for the purpose of developing broad objectives. The meeting was held in ESC's main conference room. Watkins had the vision statement taped to the wall when the members arrived.

After a brief introduction in which he explained the purpose of the meeting and cautions to observe when developing broad objectives, Watkins opened the floor for discussion. Ideas flowed surprisingly well and Watkins recorded them all. There was a definite tendency among members to propose very specific and quantifiable objectives. This wasn't where Watkins wanted things to go, but he didn't want to risk shutting off the flow of ideas by making judgmental comments.

Most of the objectives proposed contained such specifics as percentages, numbers, and dates. During the culling and refinement phases of the process, Watkins tried to encourage a general broadening of the proposed objectives, but his attempts met with little success. Consequently, ESC's final objectives were as follows:

1. ESC will increase its business volume by at least 6 percent per year.
2. ESC will increase its profits by 10 percent per year.
3. ESC will decrease customer complaints by 10 percent per year over the next five years.

Watkins felt happy to have finally accomplished something, but he wasn't sure exactly what he had accomplished. He had a vision statement, but it belonged to him, not to ESC. He had broad objectives, but they weren't really what was needed. Somehow he had to find a way to secure executive buy-in on the vision while simultaneously convincing the Steering Committee to rewrite the broad objectives. Both of these tasks were tall orders, but at least he had something to show the boss—provided he ever got in to see him.

Publicize and Communicate

At this point the Steering Committee has framed the vision, guiding principles, and broad objectives. The organization is now at the threshold of its journey into Total Quality and the attainment of world-class performance. Publicity/communication represents the first step in the implementation process that involves the entire organization. Consequently, its importance cannot be overestimated.

RATIONALE FOR PUBLICIZING AND COMMUNICATING

Rumors are a fact of life in organizations that don't keep their employees fully informed. Rumors seldom reflect reality and sometimes create problems, such as those listed in Figure 7–1. For example, by now employees have observed senior managers going to off-site meetings, cloistered as a group in several in-house meetings, and generally acting different. In such circumstances the rumor mill is likely to be operating at top speed. People will try to guess what is going on, and invariably many of the guessers will be wide of the mark. Inevitably, rumors of impending layoffs, a hostile takeover, or other dire predictions will prevail. It just seems to be human nature to expect the worst.

The positive side of this situation is that the rumor mill can be easily shut down. Simply tell people what's going on. What possible harm can it do? The rule should be,

Figure 7–1
Rumors as a Cause of
Problems

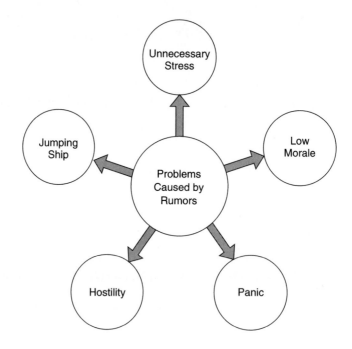

inform the organization as soon as possible. Not only will this eliminate rumors and the wasted energy that goes with them, it will also prepare people for what is about to happen. This is important, because the efforts of employees will be needed to carry out the plan.

WHY WAIT UNTIL NOW?

Why wait until Step 7 in the implementation process to start publicizing and communicating? The quick answer is that it is not wise to begin communicating about implementing Total Quality until Steering Committee members have developed the background needed to answer the questions employees will ask. By now the Steering Committee has had sufficient training in Total Quality that members should be able to respond intelligently to questions about the subject. In addition, it was only in Step 6 that the Steering Committee determined the broad objectives for the organization, something that should be done before beginning to publicize the process. Management has been observed working as a team rather than as individuals pursuing their own agendas. Seeing managers in this role helps subordinates believe that they are serious about Total Quality. In earlier steps, the Steering Committee did not have the knowledge or experience to communicate intelligently about Total Quality, or to demonstrate by action that management is serious about its implementation.

Even though it is wise to wait until Step 7 to initiate publicizing and communication initiatives, Steps 1 through 6 should not be shrouded in secrecy. In order to prevent the rumor mill from going off in all directions, the organization can be told very early

---

TOTAL QUALITY TIP

Quality Can Be Measured in Profits

"Sometimes people talk about quality as if it is some kind of abstraction, something different from the normal job. But quality is very, very real. The result of quality is profit—a wonderful measure of the kind of job we are doing for our customers."[1]

Lewis E. Platt, Chairman of the Board, Hewlett-Packard Company

---

on, via a routine newsletter article or some similar method, that the senior management is looking at Total Quality as a possibility for the organization. This should be accompanied by a brief article on Total Quality which includes anticipated benefits and success stories from other organizations. This kind of update should be released at the time the Steering Committee is formed. It should be purposely low key. At this point in the process deflecting rumors is the goal. In-depth information comes later.

EMPLOYEES NEED INFORMATION

Too many businesses unnecessarily restrict the kinds and amount of information shared with employees. The press and stockholders routinely get far more information from companies than do the employees. Yet it's the employees, not the press or shareholders, who are best able to help a business achieve its goals. The best-performing companies are typically those that make sure their employees are fully informed about all aspects of the organization's operations. Without knowing the organization's vision and broad objectives, employees do not know how best to help realize them. Departments cannot refine their efforts to work together toward a common goal unless they know what the goal is. Unless employees are fully informed, they simply cannot know whether their efforts are contributing to the overall success of the organization. If they don't know where they are going, employees cannot be expected to respond when management asks them to pedal harder. The bottom line is, informed employees are better able to align their efforts and aspirations with those of higher management. Of no less importance, they will also be in a better position to make informed personal decisions for themselves and their families. In fact, an employee without information is like a car without gas; it can't do the job expected of it.

Each employee will ultimately become involved in the implementation of Total Quality. Therefore, it will be necessary to ensure that each employee fully understands Total Quality and the organization's vision, guiding principles, and supporting objectives. These concepts may seem vague to employees, at least until they are explained and discussed thoroughly. Without this basic information and interaction, employees will not be able to maximize their efforts to help take the organization where it wants to go.

WHAT TO TELL AND HOW TO TELL IT

When it comes to communication, an organization cannot have too much. Even in organizations that attempt to communicate, there will be employees who fail to hear what is said or to read what is published. Such employees are likely to view communication in the organization as being inadequate. Communicating with employees on a large scale can be a difficult and frustrating endeavor. But despite the difficulties, organizations must communicate. The stakes are too high to do otherwise.

Given the difficulties in communicating with groups, it is important to know what to communicate. Figure 7–2 shows the six items that need to be publicized and commu-

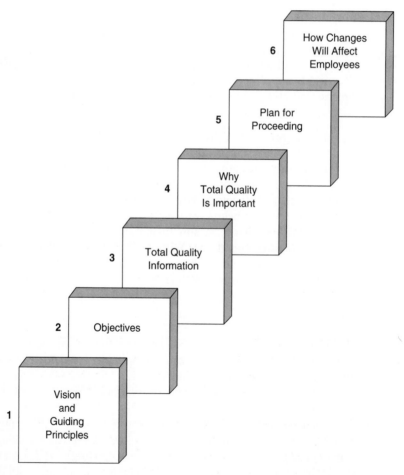

Figure 7–2
What to Publicize

nicated. The key to effective communication in an organization is to make it interesting, lively, and repetitive. Regardless of the format used, organizations should be creative and persistent.

Vision and Guiding Principles

Up to this point, the vision and guiding principles have been the proprietary knowledge of the Steering Committee. Now they must be communicated to every member of the organization, without exception. The initial introduction to the vision and guiding principles is best made in an all-employees meeting (or series of meetings, if necessary) hosted by the Steering Committee. The speaker should be the top manager. He or she should explain the vision and guiding principles and make a strong case for why they are what they are. Remember, some employees in the audience will invariably hold different views, and it is the leader's job to start converting such people immediately. The objective is to tell all employees what the vision and guiding principles are, so they can begin to internalize them, just as the Steering Committee did. This will require long-term reinforcement that should start immediately.

Although we are not big fans of posters and slogans, we do recommend that the vision and guiding principles be attractively published, framed, and posted at strategic locations throughout the organization's facilities. If properly done, posters and slogans can also be a powerful public relations strategy for the organization, since customers will also see them. The vision and guiding principles should be posted immediately after the all-employees meeting. Another effective strategy is to have the vision and guiding principles printed on attractive wallet-size cards and distributed during the initial all-employees meeting. It is not necessary for every employee to be able to recite the vision verbatim, but all should be able to paraphrase it accurately. Printed material will help achieve this level of understanding.

The publicity/communication step goes on forever. After the initial introduction, take advantage of every subsequent opportunity to discuss the vision. Keep the attention of employees focused on the vision and what they can do to help realize it.

Objectives

As were the vision and guiding principles, the broad objectives were developed by the Steering Committee as the means by which the vision will be realized. The objectives should be discussed at the all-employees meeting immediately after discussing the vision and guiding principles. A critical aspect of this discussion is establishing the connection between the vision and the broad objectives so that employees understand it. It is critical that the connection be made clear, because understanding the connection is a prerequisite of employee empowerment. Once employees understand the vision and the objectives and how they relate, their individual actions become focused and aligned. Employees who have not made the connection should not be empowered because there is no way of knowing what direction their actions will take, and their actions are likely to be counterproductive.

In this case the presenter is once again the top manager, although it can be effective to include other Steering Committee members in the discussion. This approach presents a picture of unanimity, ownership, and teamwork which may be contagious among employees.

Like the vision, the broad objectives require continuous reinforcement to keep them foremost in the minds of employees. For this reason, we suggest posting the objectives in work areas. Do this in a manner that enhances rather than clutters the work environment. In addition, take every opportunity to reiterate objectives and to discuss progress and tactics. Use the company's newsletter. If there isn't one, start a Total Quality Bulletin. Don't assume that once informed, employees will always go in the right direction. Human beings sometimes wander. It is up to the Steering Committee to keep employees on the right track. This is a never-ending task. Communicate continually.

Total Quality Information

As part of the presentation on the organization's vision, employees should have been told that it was management's intent to implement Total Quality. Some employees may know what Total Quality is, but many will not. At this point some basic information on the Total Quality philosophy and how it works should be provided to all employees. Communicate the basics in an overview. Don't make it a tutorial; employees will get in-depth training later as the teams are formed. For now, it is necessary only to introduce them to the concepts. In future communications about Total Quality (remember, this never ends), provide more insights, describe additional experiences of other organizations, and review the organization's own successes and failures. Always keep the subject in the forefront of employees' minds.

Why Total Quality Is Important to the Organization

In addition to knowing what Total Quality is, employees need to know why it is important to the organization. It is essential for the leader and the Steering Committee to spread their views of what Total Quality will bring to the organization. The best approach is to share with the employees the original thinking that went into the decision to implement Total Quality. This is usually different for every organization, but will likely include thoughts on competitiveness, customer orientation, quality focus, empowerment, involvement, and why these things are necessary if the organization is going to realize its vision and broad objectives. The leader and the Steering Committee should present this to the employees in a forum that can accommodate questions and answers. Publicity releases and newsletter articles or Total Quality Bulletins are also appropriate. We suggest using a combination of vehicles. Once the implementation starts to take hold, there will be less need to reinforce Total Quality's importance. The need will become self-evident, and that will provide the reinforcement.

How We Plan to Proceed

Employees will need to understand the organization's plan for Total Quality. The leader can begin by explaining the steps already taken and why. Next, he or she can outline the remaining steps, briefly explaining the purpose of each. It is important to highlight the fact that at Step 12 the process becomes a continuous cycle, one that eventually will become "just the way we do business."

Of particular interest to most employees will be the steps that directly involve them, such as team selection, team training, team deployment, and infrastructure change. At this point, employees should be given some information they can use. For example, they might be told that the Steering Committee will soon be looking for people for the early teams, and to let a Steering Committee member know if they would like to be considered. Be straightforward and let employees know that this is new territory for everyone, and that the approach taken might have to be modified later as lessons are learned.

A projected timetable for the various steps would also be useful for employees. If Steering Committee members are uncomfortable projecting actual dates for implementation events at this point, a tentative quarterly projection (i.e., third quarter—first team activated) is satisfactory for now. The plan is another subject that will need constant attention. As the implementation proceeds, be sure to keep the employees up to date on the latest plans, since plans have a way of changing. Use the newsletter or special Total Quality Bulletins to keep all employees informed.

How Employees Will Be Affected by Changes

The Steering Committee should let employees know that the implementation is nothing less than a total remake of the organization's culture. A change of this magnitude will not be easy for many employees. Consequently, they need to be told in advance what to expect, and that there are people on hand to help them. Employees should be given a description of what the organization is now and what it is to become in the future. Contrast current attitudes toward customers with the new customer-focused attitude. Contrast the old thoughts about what constituted acceptable quality with the new concept of customer-defined quality and continuous improvement. Contrast the old approach of trying to get employees to work harder, faster, and more carefully with the new emphasis on getting all processes used by employees under control and then continually making them better, both substantially and incrementally. Most organizations will be able to contrast a rigid, hierarchical, function-based structure and its inherent barriers to effective communication with a more flexibly structured, product- or service-oriented organization that communicates effectively and works through teams, many of which will be cross-functional. Contrast the traditional top-down flow of instructions to the self-directed work performed by teams of empowered and involved employees, with decisions made at the lowest possible level. This is also a good time to point out that strict specialization—so much a part of the traditional organization—will be an inhibitor to Total Quality. Consequently, employees will have to become more broadly skilled through training and work experience.

Some supervisors and managers will be alarmed that the organization's structure is being altered. Some will feel threatened. Even so, it is important to be completely forthright. If jobs will be eliminated, don't duck the issue. However, it is not advisable to single out individuals at this point. Point out only the classes of jobs that may be eliminated. If it is possible, and only if it can really be done, explain that displaced employees

---

TOTAL QUALITY TIP

Total Quality Is a Journey, Not a Destination

"TQM is not a destination, it is a journey. It is a cycle of constantly changing yourself to respond to internal conditions and these are natural cycles we're not paying attention to. The goal is to not always be the highest performer at a particular time; the goal is to constantly have the capacity, mindset, freedom, and the tools to continuously re-create, improve, and change."[2]

Michelle Hunt, Director, Federal Quality Institute

---

will be given other jobs. If the implementation goes well, the organization might expand enough to keep and retrain displaced employees. We recommend that this communication be done verbally by the leader. Time should be allowed for questions and answers, but the leader should not get bogged down in what jobs might be eliminated. We also recommend follow-up articles in the newsletter or Total Quality Bulletin that expand on some of these points. Consider this to be part of the Total Quality education process and keep the dialogue open.

PUBLICITY AND COMMUNICATION MEDIA

In the preceding section, we mentioned several vehicles for publicity and communication. These are summarized here and presented in Figure 7–3 as a handy reference resource.

The Verbal Presentation

This is direct communication from the speaker to the audience. Advantages of this approach include the following:

- Provides a personal touch.
- Reinforces that the speaker believes in what he or she is saying.
- Is the best approach for stirring a response.
- Audience can be involved thorough questions and answers.

Disadvantages include:

- Can become a burden for the speaker if the presentation has to be repeated several times.

Figure 7–3
Vehicles for Publicizing the
Total Quality Effort

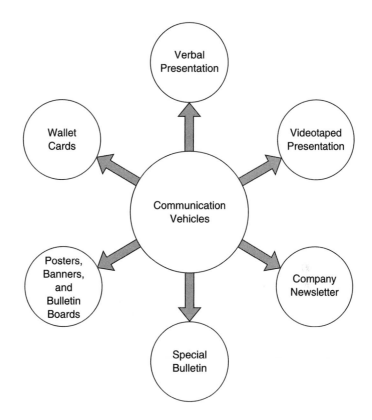

The Videotaped Presentation

This is taped communication from the speaker to the audience. Advantages of this approach include the following:

■ Has the personal touch, although not as much as a live presentation.
■ Can be edited to get it right before release.
■ Can be repeated as often as desired without burdening the speaker.
■ Can be played at the convenience of the audience.

Disadvantages include:

■ Not as convincing or believable as a live presentation.
■ Not as apt to stir a response.
■ Audience interaction is not possible.
■ Not all organizations possess the requisite equipment.

The Company Newsletter

This is written communication from the writer to the entire employee population. Newsletters normally cover a wide range of subjects of interest to the employees. Advantages of this approach include the following:

■ Wide distribution among employees.
■ Good follow-up and reinforcement potential.

Disadvantages include:

■ Space limitations, usually less than one page per article.
■ May get lost among the many other articles.
■ May not be read by all employees.

The Special Bulletin

This is a series of written bulletins about Total Quality. Advantages of this approach include the following:

■ No publication deadlines, unlike newsletters.
■ Can release at any time desired.
■ Single subject—won't get lost among other subjects.
■ Not subject to space limitations, unlike newsletters.

Disadvantages include:

■ Cost to produce and distribute.
■ Only slightly better than a newsletter as a communication medium.

Posters, Banners, Bulletin Boards

The framed vision statement, guided principles, and objectives fall into this category. They can be valuable tools if used with restraint and good taste. They should be used for reinforcement and reference, not as the primary communication vehicle. Advantages of this approach include the following:

■ Keeps message in front of employees.
■ Keeps message in front of customers; can be an effective public relations strategy.
■ Effective for reinforcing the message if used properly.

Disadvantages include:

■ After awhile, tends to blend in so that people no longer see it.

- When done poorly, contributes to clutter.
- May be relegated to "sloganeering."

Wallet Cards

The organization's vision, principles, and objectives can be printed on wallet-size cards to be carried by employees as a quick reference. Variations on this theme include the employee identification badge with this information printed on the back, and larger cards displayed on conference room tables. Advantages of this approach include the following:

- Employees like them, and when asked about the vision or objectives, will pull them out in response.
- Handy and accessible.
- Good public relations tool for customers.

Disadvantages include:

- Require frequent replacement.
- May be forgotten among other cards in wallets and purses.

The various communication vehicles explained herein by no means represent a complete list, but we think they represent the media most likely to be available for use in most organizations. However, creative companies seem to have no bounds when it comes to communicating. Some have annual Total Quality fairs, banquets, or picnics. One company even held a Total Quality cruise. All of these strategies are used to keep communication flowing and the publicity at a high level. Every organization should find ways to personalize its Total Quality publicity and communications efforts.

THE NEED FOR FEEDBACK

Up to this point, we have concentrated on communication from the leader and Steering Committee to employees. It is important to realize that there must also be communication from employees to the Steering Committee to close the loop. Without feedback from employees, the effectiveness of the organization's communication efforts cannot be determined. In addition, through their feedback, employees may be able to give the Steering Committee ideas that will make the implementation easier and/or more effective. We recommend that the organization have a system for obtaining feedback from employees as a normal, everyday function. Questions, comments, opinions, and ideas should flow among employees. The challenge is to capture feedback and put it before the Steering Committee. Every encounter with employees should be considered an opportunity for obtaining feedback. If employees don't volunteer information at first, ask questions and lead them into a discussion of Total Quality.

Earlier we discussed a Total Quality Bulletin for communications from the Steering Committee to the employees. A similar approach can be used for communicating in the opposite direction. Total Quality feedback forms can be made available throughout the organization and used by employees to ask questions of the Steering Committee, offer suggestions, or simply offer opinions. Employees should understand that the Steering Committee wants their input, welcomes it, and will use it.

SUMMARY

1. Communication gives employees the information they need to do the jobs expected of them. An employee without information is like a car without gas.
2. The following information should be publicized and communicated continuously: vision and guiding principles, objectives, Total Quality information, why Total Quality is important, the plan for proceeding, and how employees will be affected.
3. Vehicles for communicating include: verbal presentations; videotaped presentations; the company newsletter; special bulletins; posters, banners, and bulletin boards; and wallet cards.
4. Feedback from employees is essential as a mechanism for determining the effectiveness of communication.

KEY TERMS AND CONCEPTS

Banners

Bulletin boards

Communicating

Company newsletter

Feedback

Posters

Publicizing

Special bulletin

Verbal presentation

Videotaped presentation

Wallet cards

REVIEW QUESTIONS

1. Explain the rationale for publicizing and communicating.
2. Why should an organization wait until Step 7 to publicize?
3. Why do employees need to know the vision, guiding principles, objectives, and so on?
4. What information should be communicated to employees?
5. List the different vehicles that can be used for publicizing and communicating.
6. Why is employee feedback important in this step?

============ **ENDNOTES** ============

1. Lewis E. Platt, as quoted in "Report Card on TQM," *Management Review*, January 1994, 23.
2. Michelle Hunt, as quoted in "Report Card on TQM," *Management Review*, January 1994, 25.

=========== **CASE STUDY 7–1** ===========

Communicating and Publicizing at MTC

John Lee had asked each member of the Steering Committee to come to the meeting with ideas for publicizing and communicating the vision, guiding principles, objectives, and so on. After brainstorming for just over an hour, the Steering Committee had proposed the following ideas:

1. A verbal presentation by John Lee to all employees followed by a question-and-answer session in which all Steering Committee members would participate.
2. A videotaped presentation by John Lee for employees who miss the verbal presentation.
3. A Total Quality Bulletin published weekly and distributed to all employees.
4. Periodic small-group meetings conducted by individual members of the Steering Committee.
5. A Total Quality "hot line" that allows any employee to call in a question about the implementation and receive an answer within 24 hours.

The consensus was that all five ideas had merit and therefore should be pursued. After making assignments to get the ideas kicked off, John Lee adjourned the meeting confident that there would be excellent communication at MTC.

=========== **CASE STUDY 7–2** ===========

Communication Problems at ESC

Lane Watkins had a vision statement and three broad objectives. Now he needed to communicate them to all ESC employees. But there was a problem. Neither the vision nor the objectives had been officially sanctioned by executive management. To make matters worse, his repeated attempts to discuss the vision and objectives with ESC's CEO, John Hartford, had gotten him nowhere. Hartford was too busy putting out brushfires. Any discussion of such ambiguous notions as a vision and objectives would have to wait until Hartford found a way to deal with the dissatisfied customers that were either dropping ESC like a hot potato or keeping him tied up with complaints.

The message given to Lane Watkins had been simple and to the point: "I put you in charge of this Total Quality implementation, so take charge and get on with it. Do whatever you think is best."

After receiving this directive, Watkins had called a meeting of ESC's Steering Committee. The members had presented several good ideas for publicizing and communicating the vision and objectives, all of which Watkins dutifully recorded. The problem he now faced was what to do with the ideas. Neither he nor the other Steering Committee members had the authority to commit the resources needed to call an all-employee meeting, make a videotape, or publish a special weekly bulletin. John Hartford clearly had neither the time nor the patience to deal with the issue now.

As Watkins saw things, he had two options. He could commit funds he didn't control to communicating a vision that had not been officially approved, or he could put the communication ideas developed by the Steering Committee on hold for now and hope to implement them later. He didn't like either option, but in the final analysis, Watkins decided to wait. Maybe it wouldn't be much longer before John Hartford would give him some time.

Identify Strengths and Weaknesses

We now have the necessary commitment from the top and a trained Steering Committee that is operating more like a team every day. The Steering Committee has framed the organization's vision, guiding principles, and broad objectives, and communication about the Total Quality effort has begun. We are now ready for Step Eight of the implementation process: identifying the organization's strengths and weaknesses.

RATIONALE FOR DETERMINING THE ORGANIZATION'S STRENGTHS AND WEAKNESSES

It may seem odd that this step involves identifying the organization's strengths and weaknesses. After all, wouldn't any organization know that intuitively? Organizations typically know what they are good at. They are good at what they are fundamentally in business to do. For example, a real estate firm is good at selling property. A collision repair shop is good at repairing bent fenders. A hospital is good at caring for the sick and injured. We would certainly expect an organization to be proficient in its fundamental area of business. If it were weak in this area, it would quickly go out of business. But in this step we are looking for other kinds of strengths and weaknesses. We are interested in strengths and weaknesses of a more basic nature, those that affect the effectiveness of

145

the organization's processes, such as statistical expertise, data-analysis capability, problem solving, and so on.

Most key people will have their own ideas as to the organization's strengths and weaknesses, but unless the organization has made a point of collecting and coordinating lists, senior managers are not likely to have a coherent awareness of just where their organization's strengths and weaknesses lie. It is very important for the Steering Committee to have a consensus view of the organization's strengths and weaknesses, because these will become the primary considerations when making decisions on how to implement Total Quality in the organization.

There is no shortage of books on the subject of Total Quality. Many of them are excellent and provide readers with valuable information and insights. But until our companion book, *Introduction to Total Quality*, was published, little information was available on how to actually implement Total Quality. Many authors believe that since individual organizations have highly individual styles, cultures, structures, business situations, and employees, each requires a different implementation scheme. In *Introduction to Total Quality*, we took a different view. In Chapter 18 of that book we introduced the Twenty-Step Total Quality Process, and this book expands greatly on that chapter. Our view is that although every implementation must be customized in approach and detail, there are twenty fundamental steps that must be taken by any implementing organization, and those steps should be taken in a prescribed order. Step 8, in which the Steering Committee identifies the organization's strengths and weaknesses, will help the committee tailor the remaining steps to the organization's unique situation.

WHO SHOULD MAKE THE DETERMINATION?

Determining organizational strengths and weaknesses is a task for the Steering Committee. It is our recommendation that the Steering Committee perform this function in a one-day off-site meeting. At the meeting, the Committee can also accomplish Steps 9 and 10 and initiate Step 11.

Start the discussion by preparing a checklist similar to the one shown in Figure 8–1. Customize it as appropriate for your own organization. A good way to develop your checklist is through brainstorming. First, list all the skills, capabilities, and talents which are (or could be) important to your organization. Also list any exceptional departments. No discussion of specific items is permitted while the candidate list is being constructed.

Once the team has presented all its ideas, the list should be distilled to those items that are deemed especially important to the organization. List these in the Item column. The number of items will depend upon the nature of the organization, but will typically range from five to thirty. Even complex organizations are unlikely to have more than thirty. On the other hand, even the simplest organizations should have at least ten.

Next, the Steering Committee members should discuss each item to arrive at a number representing its relative strength. Rate items are on a scale of 1–5, where rating of 1 or 2 indicates a weakness and a rating of a 4 or a 5 indicates a strength. Record the

Figure 8–1
Organizational Strengths
and Weaknesses

Organizational Strengths and Weaknesses		
Item	**Weak–Strong Scale (1–5)**	**Comment**
Customer Focus–Internal	1	Internal customer concept not rooted in the organization.
Customer Focus–External	3	Most groups recognize importance of external customers.
Data Collection	3	We seem to collect lots of data.
Data Analysis	2	But we don't make good use of it.
Statistical Skills	4	A key strength.
Problem Solving	1	A serious weakness in the organization.
Exceptional Department	—	Manufacturing

rating for each item in the second column of the checklist. Don't use a 1–10 scale; it would only prolong the discussion and would not result in a more meaningful evaluation. Members must justify their ratings, partly to persuade the others and partly to justify their position, This is a good exercise in teamwork and consensus building.

Ideally, all Steering Committee members will arrive at the same number independently, but this rarely happens. Don't stop the discussion until members' rankings of an item are close, then take the average and round to the nearest whole number. For example, let's say that after discussion a member increases his or her rating from 3 to 4, and this results in four 4's and two 5's. The average is 4.33. This rating would be rounded to 4.

Applying the checklist in Figure 8–1, the organizational weaknesses in this example are Internal Customer Focus, Data Analysis, and Problem Solving. Statistical Skills is recognized as a strength, as is the Manufacturing Department, which the Steering Committee considers exceptional among the organization's departments.

TOTAL QUALITY TIP

Total Quality Must Be Customized Locally

"People tend to pick a canned program, but TQM must be unique to the culture and customers of the organization, and it must respect the company's history and where it wants to go."[1]

Jack West, President, American Society of Quality Control

TAILORING THE INITIAL IMPLEMENTATION

There are a variety of different avenues organizations can take to begin their Total Quality journey. One approach is to establish teams that are charged with solving specific problems. The implementation of a single element of Total Quality, such as Just-in-Time manufacturing, is another approach for getting started. Sometimes the key processes become the catalyst for Total Quality through teams that are dedicated to documenting the organization's existing processes, understanding them, and then seeking ways to improve them.

Armed with a locally constructed version of Figure 8–1, organizations can tailor their implementation for maximum effectiveness. The importance of this step cannot be overemphasized. It is crucial that an organization's early efforts in Total Quality be successful. Otherwise, people will become discouraged and the inevitable detractors will gain a foothold. Even though the organization can learn from its failures and be better prepared for a second attempt, it will be difficult to get people to go along the next time. The best approach is to select the initial projects very carefully so as to virtually guarantee success. There will be plenty of opportunities for more difficult undertakings once the organization has gotten the hang of the Total Quality process. Knowing the organization's strengths and weaknesses can help in steering a course that is guided by caution and rewarded by success.

TAKING ADVANTAGE OF THE ORGANIZATION'S STRENGTHS

Consider the organization represented in Figure 8–1. This organization's strengths are statistical skills and a standout Manufacturing Department. Two possibilities come to mind. With statistical skills identified as a strength, it makes sense to introduce the Total Quality effort through emphasis on its seven tools. Since these tools are applicable to every function of any organization, they could be taught to employees and promoted as a method by which problems are solved. Teams could be formed to address specific problems, and provided with the necessary training in teamwork, quality, tools, and Total Quality methods.

Statistical experts can be employed to train the teams and then monitor their performance in using the tools. Once again, the caution to pick initial problems carefully should be heeded. Also the number of teams should be limited to no more than three at first. Total Quality is not a competition to see who can field the most teams, especially in the early stages.

A department that is recognized as a strength usually is one that is well managed and in which people work together to do what is expected of them and more. It will be easier to launch something new (such as Total Quality initiatives) in a group like this than in one characterized by stress and contention. The organization in Figure 8–1 might elect to establish the first few teams in its Manufacturing Department and expand into other departments over time. Another approach would be to implement Just-in-Time programs in manufacturing as the initial effort. The use of Statistical Process Control (SPC) would also be a natural for this organization, given its strength in statistics. SPC would have to be preceded by work to characterize and understand manufacturing processes, document them, and get them in control. All of these examples would make excellent team assignments.

Most organizations will have obvious strengths that can be exploited in the early stages of implementation. The key is to pick specific projects that capitalize on the organization's known strengths, have the highest probability of success, and are meaningful. Then proceed slowly, resisting the temptation to rush into too many activities before all participants learn enough to be effective.

ADDRESSING THE ORGANIZATION'S WEAKNESSES

No organization should undertake a Total Quality implementation that requires strength where it is weak. Consequently, it is just as important to recognize an organization's weaknesses as it is to recognize its strengths. Using Figure 8–1 again, the items considered weaknesses are Internal Customer Focus, Data Analysis, and Problem Solving. Although an implementation activity based on customer satisfaction is a valid, frequently used initial strategy, we would not recommend it for this organization because the concept of the internal customer is apparently missing. Similarly, any implementation activity that requires widespread use of data-analysis techniques should be avoided, because the organization lacks the necessary skills or processes. It is difficult to avoid problem solving in any Total Quality implementation scheme, so the best approach here is to provide training in problem solving and to have both the Steering Committee and the statistical experts pay close attention to problem-solving activities to be certain they are being done correctly.

The checklist is supposed to include all characteristics that are important to the organization. It follows, then, that any weaknesses on the list must be converted to strengths. This can typically be done through training and practice, but occasionally it will be necessary to hire people with the necessary skills. Again using the example in Figure 8–1, training on the internal customer concept would be recommended. The training should do the following: stress why the concept is important, enable employees to

---

TOTAL QUALITY TIP

Total Quality Requires Skill, Knowledge, and Will

"Successful programs require three components: skill, knowledge, and will. Often the lack of will to carry out a program causes a failure. Not understanding the cause of failure also leads to erroneous conclusions about the value of the knowledge and skills to those who have the will."[2]

Ron Heidke, Director of Corporate Quality, Eastman Kodak

---

identify their internal customers and suppliers, and provide employees with the tools necessary to describe their expectations to suppliers and acquire the expectations of their internal customers. In addition, a basis should be provided through which employees are able to measure improvements in internal customer satisfaction.

Training can also be used to strengthen data-analysis capabilities. This will also involve the development of processes that facilitate analysis of the data that are readily available, perhaps through better reports or charting. This is also an area where the services of an expert will be helpful.

Whatever the weakness, if it relates to a capability that is important to the organization, it cannot be avoided for long. Left uncorrected, it will limit the organization's performance and inhibit the achievement of world-class standing.

SUMMARY

1. The rationale for determining the organization's strengths and weaknesses is that they become the primary considerations when tailoring the implementation process to the organization.
2. Organizational strengths and weaknesses should be identified by the Steering Committee. This is typically accomplished in a one-day off-site setting.
3. There are many different approaches that can be taken in the initial stages of the implementation. Regardless of the approach selected, it is critical that an organization's initial efforts be successful. Consequently, the initial implementation projects should be selected very carefully so as to virtually guarantee success.
4. Initial projects should be selected so as to take advantage of organizational strengths. Most organizations will have strengths that can be exploited in the early stages of implementation.
5. Weaknesses must be addressed as soon as possible. Training and practice will usually correct weaknesses. However, at times it will be necessary to correct a weakness by hiring the necessary talent.

KEY TERMS AND CONCEPTS

Customer focus–external	Problem solving
Customer focus–internal	Statistical skills
Data analysis	Strengths
Data collection	Tailoring implementation
Exceptional department	Weaknesses

REVIEW QUESTIONS

1. Explain the rationale for determining an organization's strengths and weaknesses as part of the implementation process.
2. Describe the process for determining an organization's strengths and weaknesses (i.e., who should do it and how?).
3. How can an organization tailor the implementation process to take advantage of its strengths and accommodate its weaknesses?
4. Explain how an organization can address its weaknesses.

ENDNOTES

1. Jack West, as quoted in "Report Card on TQM," *Management Review*, January 1994, 24.
2. Ron Heidke, as quoted in "Report Card on TQM," *Management Review*, January 1994, 25.

CASE STUDY 8–1

Identifying Strengths and Weaknesses at MTC

John Lee scheduled a full-day, off-site meeting of MTC's Steering Committee. The day was dedicated to identifying MTC's strengths and weaknesses—a sensitive but important undertaking. Steering Committee members brainstormed openly and the meeting moved along at a brisk pace. A strength identified early in the proceedings was the ability of Steering Committee members to work together.

However, after identifying this strength, Committee members began to record a long list of weaknesses, all attributable to the fact that MTC had always been a government contractor that did business on a low-bid, cost overrun basis. The following weaknesses were identified:

1. *Customer service,* because it had historically not been a major concern.
2. *Marketing,* because MTC had never concerned itself with marketing in the commercial marketplace.

3. *Purchasing,* because the low-bid, adversarial approach had always been used.
4. *Accounting,* because the department was geared toward government regulations and auditors.

After several hours of sometimes heated discussion, statistics was determined to be a strength. Engineering was also added to the list of strengths, and Manufacturing was selected as MTC's exceptional department. Manufacturing is what MTC is in business to do, and the company is pretty good at it. After reviewing the list, the Steering Committee decided that its initial implementation project would be launched in manufacturing.

CASE STUDY 8–2

Nothing But Weakness at ESC

ESC's Steering Committee met in the second-floor conference room to identify organizational strengths and weaknesses. Lane Watkins opened the meeting by explaining that he would record strengths on one flipchart and weaknesses on another. He suggested the Committee start with ESC's strengths, hoping to start the discussion on a positive note. Unfortunately his plans went awry from the start.

Nobody on the Steering Committee could identify a strength that could achieve consensus. Each member thought his or her department was strong and all the others were weak. After struggling with this attitude for over two hours, Watkins gave the Steering Committee a break in hopes of dissipating the animosity that had built up.

After the break, Watkins decided to pursue the other side of the coin. He asked participants to identify weaknesses at ESC. The first suggestion set the tone for the rest of the discussion. The Marketing Department's representative said, "This company's biggest weakness is its executive-management team." Clearly this was the consensus of the group, and it put Watkins in a difficult position. On the one hand, he didn't disagree with the group's opinion; in fact, he thought it was right on target. On the other hand, he could hardly be expected to inform the company's top managers that they were incompetent. To make matters even worse, the Steering Committee members swore each other to secrecy out of fear of retribution should their bosses learn of their opinions. Once again, the meeting broke up without much being accomplished.

Identify Advocates and Resisters

Every organization is likely to have some employees who accept change and others who resist change. We call the former advocates and the latter resisters. Advocates can be counted on to be at the forefront in the Total Quality implementation, as champions and as contributors. Resisters can be counted on to try to derail Total Quality as well as other initiatives. In Step 9 of the twenty-step process, we suggest that key people who are likely to fall into these two categories be identified.

RATIONALE FOR IDENTIFYING ADVOCATES AND RESISTERS

In the early stages of implementation, Total Quality can be fragile. Failures can undermine commitment from the top and support from the lower levels. When two or three projects fail, a typical response is to stop the implementation process while searching for the causes of the failure. The intention may be to regroup and come at Total Quality from a different angle. However, the second start inevitably will be more difficult than the first. A false start will turn marginal supporters into doubters and will fan the embers of doubt in those who opposed change from the outset. It is far better to start carefully, accounting for both the strengths and weaknesses of the organization and identifying beforehand the key people who can be counted on as advocates as well as those who are likely to be resisters. The smart approach is to make use of the advocates while working around the resisters.

IDENTIFYING ADVOCATES AND RESISTERS
AS A STEERING COMMITTEE FUNCTION

Identifying advocates and resisters is a Steering Committee function. Members of the Steering Committee should be able to identify the most influential employees in all departments. Most of these people will be in the management and professional ranks, but it is not unusual for nonmanagement personnel to exert disproportionate influence among their peers. In addition, if the organization is unionized, still another level of influence comes into play. These are the people in whom the Steering Committee should be interested.

The best approach is to list the names of all employees, regardless of level, who are influential. Then, discuss each person on the list to determine his or her likely stance relative to Total Quality. Persons whose stance is predicted to be negative are considered likely resisters. This process is not intended to stigmatize anyone; Steering Committee members must be careful to ensure that this doesn't happen.

The process is hardly scientific, in that it involves making an educated guess as to how people will react to Total Quality implementation. Hence, it is not necessarily accurate, and it is therefore important that people not be treated differently as a result of the process. Employees are human beings, and human beings can be unpredictable. Perceived resisters may turn out to be champions, and perceived advocates may resist the hardest. Even so, the identification process is a useful endeavor that can increase the likelihood of success if the Steering Committee uses potential advocates effectively while protecting initial projects from potential resisters.

Keep in mind that the advocate/resister issue is temporary. As Total Quality begins to catch on as a result of successes, resisters tend to become advocates, or at least become less negative. In organizations that have had two or three years of success in Total Quality, it is difficult to find detractors.

Once the organizational strengths and weaknesses are determined, identifying the likely advocates and resisters is a natural follow-up activity. Both of these tasks usually can be accomplished in one day. We recommend that the Steering Committee make the identification of advocates and resisters a second part of the off-site meeting discussed in Step 8. In organizations that conduct formal performance appraisals, the Steering Committee may be able to obtain information that might shed light on the characteristics of their employees.

CHARACTERISTICS OF ADVOCATES AND RESISTERS

Some typical characteristics of advocates are illustrated in Figure 9–1. Employees who can be expected to support the Total Quality implementation can typically be described using one or more of these characteristics.

■ *Innovator:* Likes to try new methods and new ideas concerning how work is done.
■ *Can-Do Attitude:* Has a positive attitude toward work and new challenges.

Figure 9–1
Typical Characteristics of
Advocates

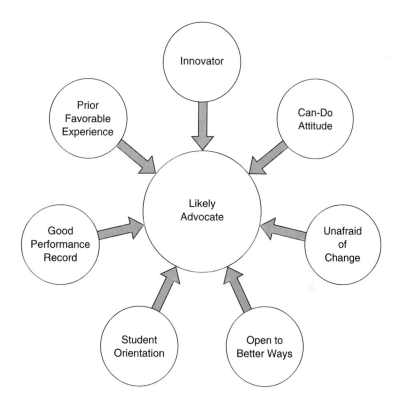

- *Unafraid of Change:* Has little or no fear of change and sometimes seems happiest in the midst of change.
- *Open to Better Ways:* Is willing to try ideas proposed by others.
- *Student Orientation:* Enjoys self-study, seminars, workshops, and school.
- *Good Performance Record:* Consistently performs well in all duties.
- *Prior Favorable Experience:* Has worked in another organization that successfully implemented Total Quality.

On the list of influential employees, categorize those who demonstrate any of these characteristics as potential advocates.

Some typical characteristics of resisters are illustrated in Figure 9–2. People who possess one or more of these characteristics may be expected to be resisters, at least in the initial stages (i.e., before the implementation has had a chance to demonstrate success).

- *Can't-Do Attitude:* Spends more time and energy trying to convince others that something cannot be done than it would take to do it in the first place.
- *Prefers Status Quo:* Takes comfort in the predictable sameness of the current way of doing things.

Figure 9–2
Typical Characteristics of
Resisters

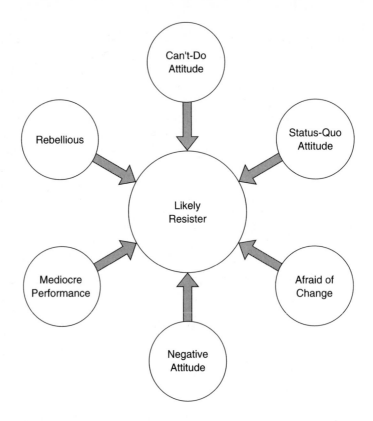

- *Afraid of Change:* Fearful of change, either because it represents the unknown or because of a fear that a new system or culture will result in loss of status, power, or influence.
- *Negative Attitude:* Initial reaction to anything is usually negative.
- *Mediocre Performance:* Job performance is less than satisfactory.
- *Rebellious:* Can be counted on to work against anything and everything.

On the list of influential employees, classify those who possess some of these characteristics as likely resisters. Do not involve them in the early Total Quality initiatives. After a few successful efforts have been completed, these initial resisters will no longer pose a threat. Some may even begin to convert.

ADVOCATES AND RESISTERS IN THE INITIAL STAGES OF TOTAL QUALITY

Employees who were identified as the most likely advocates should be involved in the initial implementation process. In Step 12, when the organization's tailored implementation plan is developed, the information used to identify the advocates will be used again to make sure some of these employees are involved in initial projects as appropriate. The list of likely resisters will be used to make sure that potential resisters are not assigned to the initial teams, and that the initial projects fall in areas in which resisters have little or no influence. Like the information on strengths and weaknesses identified in Step 8, the list of advocates and resisters is intended to increase the probability of success in the early stages, thereby giving Total Quality a better chance to gain a foothold.

During the initial stages of implementation what should be done about resisters? At this point, the answer is: nothing. Remember, you might be wrong about a given individual. Developing the list of advocates and resisters was not intended to stigmatize anyone. Over time, the true attitudes of employees will become apparent. In the meantime, if some employees deliberately obstruct the implementation, management judgment must be exercised. Clearly, no organization can tolerate behavior that is contrary to its vision and principles. However, in our experience, the real detractors typically either change their attitudes or leave on their own.

SUMMARY

1. It is critical to identify potential advocates and resisters early in the implementation process so that this information can be used to help ensure the success of initial projects.
2. Identifying likely advocates and resisters is the job of the Steering Committee. However, the process should not be allowed to stigmatize employees.
3. Likely advocates tend to have one or more of the following characteristics: innovator, can-do attitude, unafraid of change, open to better ways, student orientation, good performance record, and prior favorable experience with Total Quality.
4. Likely resisters tend to have one or more of the following characteristics: can't-do attitude, prefers status quo, afraid of change, negative attitude, mediocre performance, and rebellious.
5. Advocates should be involved in the early stages of implementation; resisters should be excluded. Once the organization has experienced several successful initiatives, the resisters will be less of a problem.

KEY TERMS AND CONCEPTS

Advocate

Afraid of change

Can-do attitude

Can't-do attitude

Good performance record

Innovator

Mediocre performance

Negative attitude

Open to better ways

Prefers status quo

Prior favorable experience

Rebellious

Resister

Student orientation

Unafraid of change

REVIEW QUESTIONS

1. Explain the rationale for identifying advocates and resisters.
2. What is the inherent danger in identifying employees as likely advocates and resisters? What can Steering Committee members do to overcome this inherent danger?
3. List the typical characteristics of likely advocates of Total Quality.
4. List the typical characteristics of likely resisters of Total Quality.
5. Explain how to take advantage of the list of likely advocates and resisters during the initial stages of implementation.

CASE STUDY 9–1

Advocates and Resisters at MTC

Once the Steering Committee had identified MTC's strengths and weaknesses, John Lee turned the discussion to the subject of likely advocates and resisters. However, before recording any names, Lee cautioned participants by saying, "What we are about to do is a dangerous undertaking. The temptation will be to write off the employees on the list of resisters. You and I have got to make a special effort to see that this doesn't happen. I want to see that every employee, regardless of his or her initial attitude, is given a fair chance to make the team. Remember, we are guessing here. This is new territory for all of us. We could guess wrong."

Having cautioned participants about stigmatizing likely resisters, Lee asked for names. After an hour the Steering Committee had developed two lists, one containing the names of employees likely to support the implementation and the other containing the names of likely resisters. Only the Steering Committee members would have access to the lists which, for the sake of confidentiality, were called Group A (advocates) and Group B (resisters).

CASE STUDY 9–2

Resisters and Resistance at ESC

Lane Watkins called a special meeting of the Steering Committee to discuss potential advocates and resisters. After the fiasco of the last meeting on ESC's strengths and weaknesses, Watkins was determined to accomplish something tangible. He began the meeting with a well-rehearsed cautionary statement about the importance of identifying likely advocates and resisters while at the same time taking care to ensure that resisters are not treated differently.

Watkins handed out copies of a list of typical characteristics of advocates and resisters. Then he handed out a list of ESC's employees and asked all participants to code each name with an "A" or an "R." Once all Steering Committee members had their lists coded, Watkins planned to lead them in a discussion that would result, he hoped, in a consensus about each employee.

At first, it looked as if the process was working. All participants appeared to be mentally comparing employees against the two sets of criteria. Watkins felt his spirits lifting. "Finally," he thought to himself, "we are going to make some progress!" After about 30 minutes the group appeared ready to discuss their individual lists.

"Who would like to begin?" Lane Watkins asked confidently. The Marketing Department's representative raised his hand and Watkins gave him the floor. "Lane, I have coded my list as follows: In the category of likely advocates, I have one name: Lane Watkins. In the category of likely resisters I have the entire executive management team. As to the rest, I categorize them as irrelevant as long as our CEO and VP's aren't on board."

As the other participants read their lists, it became clear that Watkins's feelings of confidence had been premature. The only consensus that would be reached in this group was that ESC's biggest problems were at the top.

Establish a Baseline for Employee Attitudes and Satisfaction

Some organizations pay close attention to their employees' attitudes and level of job satisfaction. Others are less concerned about this issue. Unfortunately, too many organizations have an attitude toward employees that is characterized by the following statement: "We gave you the job; whether you like it or not is your problem." Such an attitude toward employees is completely out of step with the Total Quality approach to doing business. All managers should recognize the fact that employees who are generally satisfied with their jobs and who enjoy their work are more likely to perform better than those who are dissatisfied and dislike their work.

RATIONALE FOR DETERMINING EMPLOYEE ATTITUDES AND JOB SATISFACTION

When an organization begins a Total Quality implementation, employees typically react to the changes in a positive manner. This is because they become more involved in the decisions and operation of the organization, at least to the point of having more influence over the processes with which they are involved. When this happens, employees tend to develop a heightened interest in their work and begin to feel a sense of owner-

161

TOTAL QUALITY TIP

The Implementation Factor in All Three Angles

"If TQM is an equilateral triangle, companies must factor in all three angles: management leadership, employee involvement, and technical systems. The problem is, companies don't do all three, they only do one or two."[1]

David Gregerson, Vice President for Quality Carrier Corp.

---

ship and power not previously enjoyed. Involvement and ownership lead to enhanced job satisfaction. An organization needs to be able to measure such changes. It is important to have hard data rather than a vague impression of prevailing attitudes and levels of job satisfaction. By clearly identifying current attitudes and levels of job satisfaction, an organization can establish a baseline against which to measure future gains. In other words, the primary reason for establishing a baseline is to identify the starting point for making improvements.

If your organization has ignored employee attitudes in the past, another reason for establishing a baseline is to bring home the point that employee attitudes and job satisfaction are important in a Total Quality environment. In such an environment, it is up to management not only to measure attitudes on a periodic basis, but to determine the causes for poor attitudes and attempt to eliminate those causes. Consider the analogy of a machine that becomes temperamental, requiring more and more attention and adjustment in order to maintain its standard output. In such a case, management is likely to be quick to investigate, determine the cause, and implement the necessary repairs. However, when employees develop bad attitudes, lose interest in their work, and are generally unhappy, management is less likely to investigate, identify, and repair. A common attitude is, "He's either going to have to straighten himself out, or we're going to let him go."

Sometimes the cause of an attitude problem has nothing whatever to do with work, but lies within the employee. However, it is also possible—and quite common—that employees' bad attitudes and low level of job satisfaction are rooted in the behavior of management. When this is the case, employees, like machines, require attention and adjustment in order to maintain their standard output. However, employees with bad attitudes and low job satisfaction require more effort on the part of management than do cranky machines. Further, no amount of talking, cajoling, or even threatening will ever fully return an unhappy employee to a higher level of performance. The way to obtain the best from employees is to eliminate the root causes of their dissatisfaction, give them a voice in how their jobs are to be done, and allow them to take pride in their work. If this is the first time the organization has tried to gauge employee attitudes and job satisfaction, the process should be standardized and used continually.

DETERMINING ATTITUDES AND JOB SATISFACTION AS A STEERING COMMITTEE FUNCTION

Determining employee attitudes and levels of job satisfaction is a Steering Committee function. It should actually be assigned to an augmented Steering Committee, since the Steering Committee may not have the necessary expertise. In this case, help should be secured from a facilitator or consultant. Methods for determining employee attitudes are covered more fully in the next section. In this section, the point is that the responsibility to make the determination, regardless of how it is done, rests with the Steering Committee.

METHODS FOR DETERMINING EMPLOYEE ATTITUDES AND JOB SATISFACTION

Employee attitudes and job satisfaction levels may be determined in a number of ways. Large organizations sometimes employ firms that specialize in this task. The most common method is to conduct a survey by having all employees complete a questionnaire. Such questionnaires typically consist of a list of statements with which the employee may agree or disagree. The extent of the employee's agreement or disagreement with each statement is indicated by ranking the item on a scale of 1–5 or 1–10. The list of statements is designed to elicit the true feelings of respondents, and will have been modified over time so as to result in scientifically accurate, statistically sound data. Survey firms typically provide the results in compiled form that can be sorted in a number of ways: by department, by hourly workers, by salaried workers, and so on. For the most accurate view of employee attitudes and job satisfaction, this is the best approach. Disadvantages of this approach are that such surveys can be expensive, must be scheduled 30–60 days in advance, and require another 30–45 days for the compilation of results and production of a report.

Another approach is for the organization to perform its own surveys. A Human Resources person with appropriate training can develop an effective questionnaire, or a consultant can be employed to develop the instrument. Figure 10–1 is a generic survey instrument that can be tailored to meet the requirements of most organizations.

Regardless of the instrument used, what is important is that it reflect in general terms how employees feel about their jobs, how the organization is being run, and whether employees think they have a vital role to play in the future of the organization. It is essential that the survey cover all aspects of the operation, from communication to recognition, from working conditions to management style, from employee involvement to processing of complaints. It is also important that objectivity be built into the survey instrument, and, above all, that anonymity be assured. The most serious drawback to an organization's developing and conducting its own survey is the potential for the results to be invalid because the survey design was not sound, or for the survey to have a negative effect on employees because it was improperly prepared or conducted.

Another approach to identifying employees' attitudes involves no formal survey instrument. Rather, the Steering Committee, augmented by either an in-house expert or

Please read the following statement and check the box at the end of the statement that most closely expresses your feelings about the statement. Box 1 indicates Strong Disagreement; Box 2 indicates Some Disagreement; Box 3 is for Neither Agreement nor Disagreement; Box 4 indicates Some Agreement; and Box 5 indicates Strong Agreement with the statement.

Statement	Disagree			Agree	
	1	2	3	4	5
1. I enjoy working here.					
2. I feel that my job is important.					
3. If I do an outstanding job, I will be recognized for it.					
4. The company provides the tools I need.					
5. The company provides the information I need to do my job.					
6. Managers are helpful and supportive.					
7. I am proud of the work we do here.					
8. If I have something to say, management will listen.					
9. Management keeps us informed of what is going on and how the company is doing.					
10. Working conditions in my area are good.					
11. Our managers are friendly and open.					
12. Our managers are competent in their jobs.					
13. I am not afraid to ask for help or advice on my job.					
14. If my manager doesn't know something, he/she will admit it.					
15. I feel that I am paid fairly.					
16. Managers here get special treatment.					
17. I feel free to make suggestions.					
18. Employee suggestions are taken seriously.					
19. Management encourages us to think.					
20. We all try hard to please our customers.					
21. Teamwork is encouraged here.					
22. If I make a mistake, I will not be punished.					
23. Our jobs have written procedures.					
24. Our procedures are good.					

Figure 10–1
Employee Attitude and Job Satisfaction Survey

a consultant serving in a facilitating role, spends several hours in an objective discussion of the issue, using the individual viewpoints of the members as the data. It is possible to develop a reasonably accurate view of employee attitudes and job satisfaction using this approach. It will not be as meaningful as the employee survey, but it does have the advantage of being unobtrusive.

The limitations of this approach include the potential for biased, second-hand data and a lower probability of identifying root causes of specific problems. If the discussion approach is used, keep in mind that what is being identified is a baseline or benchmark of current attitudes and levels of job satisfaction from which to measure future progress. Using a scale allows the Steering Committee to establish a numerical baseline against which future measurements can be compared. The process can be repeated in six months to a year to determine whether attitudes have improved over the benchmark—for example, from a "2" to a "3". This amount of change will be easy to detect, and implementing Total Quality over a six- to twelve-month period should produce a noticeable change.

USING EMPLOYEE ATTITUDE/SATISFACTION INFORMATION

The information gained from determining employee attitudes and job satisfaction is used primarily as a baseline from which to measure improvement. If the information is obtained using one of the survey methods, it will also be invaluable for pointing out areas that need attention, either through direct management action or as projects of special improvement teams. For example, if statements 5, 23, and 24 in Figure 10–1 receive low ratings by employees, the organization is probably weak in the area of procedures or work instructions. As part of the implementation of Total Quality, teams could be assigned to develop better procedures for key processes.

REPEATING THE PROCESS

If the first determination of employee attitudes and job satisfaction is made with the intention of using it as a baseline, it follows that subsequent determinations will have to be made. The question is, how frequently? We suggest that the process be repeated annually for the first two years of the implementation. Circumstances change fast as a result of Total Quality, and the Steering Committee needs to be aware of the degree to which employees are encouraged or discouraged by the changes. We suggest that the same survey instrument be used in subsequent administrations so that any change in results reflects changes in the organization rather than changes in the instrument.

In addition to repeating the survey using the same instrument, the Steering Committee should be sensitive to indicators of changes in attitudes and job satisfaction during the critical first six to twelve months of implementation. As each new Total Quality initiative is implemented, look for evidence of change. If the change is positive, the organization may want to do more of the same. If the change associated with a Total Quality initiative is deterioration in employee attitudes or job satisfaction, determine the cause

---

TOTAL QUALITY TIP

Total Quality Isn't Just a Set of Tools; It's a Management Philosophy

"TQM is not simply a set of tools applied by teams but a radically different system of management based on a philosophy of prevention, management by fact, employee satisfaction and growth."[2]

Christopher W. Hart, President, TQM Group

---

immediately and take appropriate corrective measures. A Total Quality organization always considers employee attitudes and job satisfaction to be key factors for success. Consequently, these factors should be improved continuously and forever.

A WORD OF CAUTION

If employee attitudes and job satisfaction are going to be determined using a questionnaire, the cooperation and participation of the employees will be critical. Prepare employees by telling them in advance that the data will be collected, what they will be asked to do, and why the Steering Committee is interested in the data. Assure the employees that the information they provide will be confidential, that it cannot be traced to any individual, and that it will never be used against them. Be sure to let employees know that it is important for them and for the organization that their responses be honest, and that they not be influenced by what they think management might wish to hear. Some organizations have asked employees to take the survey instrument home and complete it on their own time. We consider this approach unreasonable, and so will the employees. Have the instruments completed on company time after providing a few minutes of explanation and instruction. These precautions will help ensure that the data accurately reflect reality.

SUMMARY

1. The primary reason for determining employee attitudes and job satisfaction is to establish a baseline that serves as a starting point for making improvements.
2. Identifying employee attitudes and job satisfaction levels is the responsibility of the Steering Committee.
3. Three basic approaches can be used to measure employee attitudes and levels of job satisfaction: a survey designed and conducted by an outside consultant, a survey designed and conducted internally, and discussion and consensus by the Steering Committee.

4. The baseline for employee attitudes and job satisfaction levels is used as a starting point for future improvement. In addition, the results of the employee survey can be used for giving assignments to improvement teams.
5. The survey should be repeated periodically to identify changes in attitudes that occur as a result of the implementation of Total Quality. When attitudes change for the worse, immediate corrective action should be taken.
6. It is important that employees understand that the information collected in an attitude survey is confidential and will not be used against them. It is critical for both the organization and employees that honest answers are given.

KEY TERMS AND CONCEPTS

Baseline   Employee attitudes
Benchmark   Job satisfaction
Confidential   Survey instrument

REVIEW QUESTIONS

1. Explain the rationale for determining employee attitudes and job satisfaction.
2. Why is determining employee attitudes a Steering Committee function?
3. Describe three approaches for determining employee attitudes and job satisfaction.
4. How should the Steering Committee use the information collected from an employee attitude survey?
5. Once a baseline has been established, how often should the employee attitude survey be repeated?
6. Why is confidentiality important when conducting employee attitude surveys?

ENDNOTES

1. David Gregerson, as quoted in "Report Card on TQM," *Management Review,* January 1994, 23.
2. Christopher W. Hart, as quoted in "Report Card of TQM," *Management Review,* January, 1994, 24.

CASE STUDY 10–1

Determining Employee Attitudes at MTC

John Lee decided to ask MTC's outside facilitator to develop a survey instrument for measuring employee attitudes. When the task had been accomplished, Lee called a companywide meeting to announce that the survey was about to be undertaken. He explained that the survey's purpose was to establish a baseline against which improvements could be compared. He assured employees that all information collected would be confidential; no one would know who filled out the specific survey form. He then explained that the results would be tabulated and summarized, and that he and other members of the Steering Committee would have access to the summary only.

With the purpose of the surveys stated and the confidentiality of feedback explained, Lee began a question-and-answer session during which he stressed the need for honest answers.

There were several questions about what types of information would be asked for and how it would be used. Lee read several questions from the final survey instrument and explained that all information collected would be used to make improvements against an established baseline.

The survey was completed over a three-day period beginning the next day. The results established a baseline that was already respectably high; however, it did show room for improvement in all areas. The Steering Committee used this information to get three improvement teams started right away. Immediate attention was needed to improve written procedures, employee recognition, and the employee-suggestion system.

CASE STUDY 10–2

Attitude Problems at ESC

Lane Watkins knew he had a tiger by the tail if he was going to measure employee attitudes at ESC. They weren't just bad, there were hostile. He knew that already. But, on the other hand, maybe an employee attitude survey would finally get the attention of higher management. Thinking, "The most I have to lose is my job," Watkins decided to move forward with the survey.

He called a meeting of ESC's Steering Committee and explained what needed to be done. The collective response of the members was brief and to the point. "Good luck. But leave us out of it." Feeling even more than usual like a man adrift in a sinking rowboat, Watkins decided to move forward anyway.

Borrowing liberally from one of his old college textbooks, Watkins developed a one-page survey instrument and distributed it to all departments. All employees at ESC except higher management received a survey form. Within a week almost all of the forms had been returned and Watkins faced the daunting challenge of summarizing and analyzing the data. This he accomplished in the period of one weekend, during which he barricaded himself in his apartment and worked around the clock, stopping only for a few well-deserved catnaps. By late Sunday night, Watkins had the results in hand. Now the question was, what to do with them?

The results were just as bad as Watkins had predicted. After contemplating his next move for several days, Watkins decided to cross the Rubicon. He drafted a brief memorandum to ESC's CEO, John Hartford, and attached the summary of the attitude survey. Then he sat back and waited for the explosion.

Establish a Baseline for Customer Satisfaction

Most organizations agree that it is helpful to know whether their customers are satisfied with their product or service. As long as there is competition, this is generally true. However, where competition is limited or nonexistent, organizations tend to lose sight of the importance of customer satisfaction. A government agency is one example of an organization that provides a service to customers yet seldom recognizes them as such. Of course, there are exceptions. Surprisingly, the Internal Revenue Service is one exception. The IRS is well on its way to becoming a Total Quality organization. Another exception is the Armed Services, which is also moving down the road to Total Quality.

Unfortunately, many governmental entities are still where they have been for years: either failing to recognize that their job is to satisfy their customers (the public), or worse yet, simply not caring whether their customers are satisfied. Although we cite government agencies as an example, such behavior is not limited to the government. Examples can also be found in the private sector. You have no doubt encountered blatant disregard for customer satisfaction on a number of occasions, perhaps in restaurants, stores, automobile repair shops, and other businesses. Using Figure 11–1, see if you can recall occasions when you, as a customer, were less than satisfied with an organization's product or service.

Suppose you checked into a hotel and found that the television was out of order and the towels in the bathroom were dirty. What would you do? Chances are you would

Figure 11–1
All Types of Businesses Can
Fail to Satisfy Customers

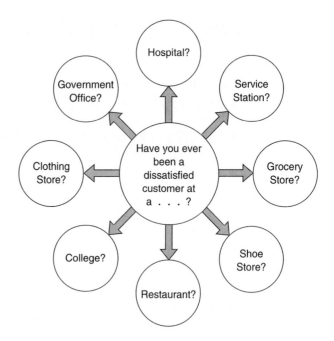

notify the desk clerk of the situation right away, and you would expect the desk clerk to have your problem resolved expeditiously. You would not attempt to locate a television repair technician, nor would you call the housekeeper to ask for clean towels. The desk clerk is the customer's primary point of contact in the hotel, and consequently is the logical person to contact with any hotel-related problem. Unfortunately, employees don't always see their responsibilities in the same way their customers see them.

A classic example of an employee who is either ignorant of or indifferent to the concept of customer satisfaction is found in the case of the hotel desk clerk who wrote to "Dear Abby" on behalf of front desk clerks throughout the world. His purpose was to let the public know that he and his colleagues do not repair televisions, unstop toilets, or change light bulbs. Nor do they, according to this clerk, deliver towels, carry luggage, or run errands. He went on to explain that a desk clerk is paid to greet and register guests and to make sure they pay and turn in their keys on the way out.

From a customer-satisfaction perspective, this person is the desk clerk from hell. Abby's response to the writer was brief, to the point, and accurate. She explained to the desk clerk that his job is *customer satisfaction*—period.

RATIONALE FOR DETERMINING CUSTOMER SATISFACTION

The primary rationale for determining customer satisfaction is survival. An organization that fails to satisfy its customers will not stay in business very long.

Customer satisfaction is the very essence of Total Quality. It is impossible to become a Total Quality organization without understanding and applying the following concepts:[1]

■ Customers are necessary for all enterprises, public or private. If there are no customers, there is no need for the enterprise.

■ Satisfied customers not only bring return business, they represent a powerful positive advertising force.

■ Dissatisfied customers not only become customers for the competition, they represent a powerful negative advertising force.

■ Even for enterprises that have limited competition, dissatisfied customers ultimately will find some form of recourse.

■ The customers, and only the customers, can pass final judgment on the quality and value of products or services.

■ Internal customers should command the same respect as external customers in terms of satisfaction.

■ When an enterprise lists its major objectives, total customer satisfaction must be at the top of the list.

Customer satisfaction is at the heart of Total Quality, but as a concept it has been around much longer than the Total Quality movement. As Total Quality has evolved, it became clear that only by putting the customer first will an organization be poised for the changes necessary to become a world-class performer. Putting the customer first is a kind of permanent catalyst for improvement throughout the organization.

On a more pragmatic level, rational managers agree that it is better to have satisfied customers than dissatisfied customers. However, many organizations are unable to start with more than an intuitive feeling about their level of customer satisfaction. There are several indicators, such as profit trends, repeat customers, and new customer trends, that can be checked. However, indicators are easy to disregard or rationalize. To really know where you stand in terms of customer satisfaction, a special effort must be made to solicit information directly from customers.

In order to measure the degree of success or failure of a change, the starting point must be known with certainty. Hence, before changing anything, an organization should first measure customer satisfaction. This initial measurement of customer satisfaction information establishes the baseline from which improvement or deterioration can be reliably measured.

When customer satisfaction is reassessed after six months or a year, if there is solid evidence that customer satisfaction has improved, the Total Quality implementation process has been validated. If definite improvement is not evident, then the implementation should be analyzed to see why one of its primary benefits–customer satisfaction–is not being achieved.

APPROACHES TO DETERMINING CUSTOMER SATISFACTION

One of W. Edwards Deming's absolute rules was that managers must base decisions on facts. This means that the initial estimate of customer satisfaction must be based on factual data rather than intuition or guessing. Organizations also need to know how far and in what direction they went as a result of the change. There is only one source for factual data on customer satisfaction: the customer. The question, then, is not *where* to find the information, but *how* to find it. A number of techniques for obtaining data on customer satisfaction follow.

■ The Customer Service Department can be a gold mine of customer-satisfaction data. The problem with this source, however, is that virtually all contacts with the Customer Service Department are negative. Satisfied customers don't usually call customer service. Even so, this negative data can give you a good idea of the kinds of problems customers are experiencing with your products or services. This information can help you determine where to assign resources, not only to improve customer satisfaction, but also to make fundamental improvements in products and services.

■ If your customers have a user association, it can provide information about the kinds of problems customers have. Again, this information can be useful in pointing out areas in need of improvement.

■ Some organizations regularly meet with customer panels set up expressly to provide feedback. Some organizations mail surveys to members of a customer panel. In either case, such panels have worked well in making the organization aware of issues of interest to customers. The key to having successful customer panels is the selection process that creates the panel. Care must be taken to make the panel reflect the demographics of the entire customer base. Another important factor is the meeting agenda. If the object of the meeting is to find out how customers feel about the organization and its products and services, a survey must somehow be integrated into the agenda. As with any survey, the results will be no better than the survey itself, so take the time to develop a valid instrument. The advantage of customer panels is that the panel members are known to be willing participants, and hence tend to be more interested and less time-constrained than customers selected at random for a direct survey.

■ The direct survey is the most commonly used method for determining customer satisfaction. Surveys can be conducted either by mail or by phone. With mail surveys, the customer completes and returns a survey form. With phone surveys, customers are interviewed directly and the interviewer records the customer's response. In both cases, it is important that the surveying organization be sensitive to the possibility of imposing on respondents. The survey should be brief, and the same customer should not be surveyed repeatedly unless that customer is a panel member.

If your organization has never attempted to assess customer satisfaction, we suggest the direct customer survey as the one most likely to produce the desired results. If your organization does not have the internal capability to develop a survey instrument or to conduct the survey, consulting firms are available to help.

Whether the survey is conducted by telephone or mail, remember that it places a burden on customers. Make it as easy on them as possible. Limit the number of questions to between six and ten and make each question count. Provide the means to simply check a box rather than requiring narrative answers. Further, since you are limited to just a few questions, focus on the fundamentals of your business and its products and/or services, not on tangential issues. For example, ask if your product met the customer's expectations, not whether he or she liked some seldom-used feature. Survey questions should be geared toward the customer's interests. Customers are typically more interested in the basics than in the less critical elements. If a selling process is involved, the customers will be interested in events surrounding the purchase, such as whether they felt comfortable or intimidated or pressured. The customer will also be interested in what happened after the sale. If prepared properly, the survey instrument will be viewed by the customer as evidence that he or she was not forgotten after the sale.

A key point you should take from this section is that a survey form or a telephone interview should be brief. One that takes ten or fifteen minutes to complete is too long. You should be able to collect the information needed in one or two minutes; any longer, and the customer will lose interest. Yankelovich Partners, a leading market-research and public-opinion survey firm, finds that customer attention falls off after only one minute.[2] So keep it short, keep it basic, and make it interesting.

DETERMINING INTERNAL-CUSTOMER SATISFACTION

Up to this point we have emphasized the satisfaction of external customers. Everything discussed so far also applies to determining internal customers. Every organization has internal customers. In fact, every employee is a customer to some other employee, who in turn is a customer to another employee. For example, the Purchasing Department's internal customers are those who use the materials, goods, and services procured by the Purchasing Department. Departments within the organization can learn from their internal customers in the same way the whole organization learns from its external customers.

Internal customers can be surveyed in a manner similar to that used to survey external customers. Internal surveys should be handled on a face-to-face basis if possible. Beyond this, the same rules as those for surveying external customers should be observed. The department conducting the survey can use the data collected as a benchmark for current internal-customer satisfaction, and can measure the results of any changes made in the future against that benchmark. If improvements have been made as a result of a change, keep going. If a change has caused a decrease in internal-customer satisfaction, as measured against the benchmark, then find out what went wrong, correct it, and measure again. This technique can be applied to every department, just as it can to the total organization.

OBTAINING MEANINGFUL INFORMATION

The obvious use of customer-satisfaction data is establishing a baseline against which to measure future improvement. This means the data must be quantifiable. When measuring customer satisfaction, consistency is important. In other words, each successive time you survey customers, the information sought and the way it is quantified should be consistent with the initial survey. Otherwise, you could end up comparing apples to

---

TOTAL QUALITY TIP

Organizations Should Listen to What They Don't Want to Hear

"The foundation of superlative performance rests on being self critical and asking others to comment on our performance. This means listening to what we do not want to hear."[3]

Philip E. Atkinson

---

oranges, a situation which precludes meaningful interpretation of the data. We will not attempt to suggest either the questions or the quantifying procedure here, because the surveys used by different organizations will necessarily vary widely. Nevertheless, some fundamental techniques should be observed:

■ Use questions that get at the basics of your products or services.

■ If appropriate for your organization, cover the total customer experience, from the sale, to use of product or service, to what happens after the sale.

■ Weigh the values of the questions by importance as perceived by the respondents. The survey instructions should ask the customer to "rank the following questions 1 (the most important) through 8 (the least important) in terms of significance for you." When the survey data have been collected, make a composite ranking based on customer inputs. The ranking tells you what's important to your customers and serves as the basis for the weights you assign.

Customer-satisfaction surveys have a secondary value that is also important to the organization. Every organization should take advantage of the fact that its customers are revealing what they like or don't like and what is important to them. This information provides what W. Edwards Deming called *profound knowledge*, or knowledge from outside. Dr. Deming believed that an organization is incapable of understanding itself. It must have an outside view to evaluate present actions and policies. The outside view reveals what must be done to improve, or at least what needs to be improved. By focusing on what is important to the customer, resources can be used to their greatest advantage and to the fullest benefit of the organization, customers, and employees.

SHORT-TERM AND LONG-TERM FOLLOW-UP

New surveys should be conducted at regular intervals. Some companies sample customers monthly, some quarterly, and others annually. We suggest that until the Total

Quality initiative is at least several months old, there is probably no point in conducting a second survey. A good time to conduct the second survey is twelve months after the first. However, if a significant change has occurred that should have impacted customer satisfaction, the second survey can be conducted sooner.

After the second survey has been conducted, we suggest conducting continual surveys at quarterly or monthly intervals. A good rule of thumb is that a monthly survey should represent one-twelfth of an annual survey, but should not be so small that it loses statistical validity. Surveying monthly or quarterly minimizes the burden on your staff. If you contract with an outside survey firm, once-a-year surveys are usually more economical. The key is to conduct surveys frequently enough to keep you informed of progress—or the lack thereof.

CUSTOMER SATISFACTION VERSUS CUSTOMER DELIGHT

Once customers are clearly satisfied, the organization can take the next step: achieving *customer delight.* This is where world-class organizations are headed, and to have a competitive advantage, this is where your organization must go. How can you tell the difference between satisfied customers and delighted customers? Satisfied customers are reasonably happy with their purchase, but will still consider the product or service of competitors when making their next purchase. Delighted customers are so happy with their relationship with your organization that they won't even consider purchasing from the competition.

=========== SUMMARY ============

1. The rationale for determining customer satisfaction is survival. Organizations live or die depending on how well they satisfy their customers. If a customer is satisfied, an organization needs to know why. If a customer is not satisfied, an organization needs to know what to improve.
2. Customer-satisfaction information can be collected through mail and/or telephone surveys. Using customer panels can improve the quality of the feedback received. Organizations should solicit feedback from both external and internal customers.
3. Initial customer-satisfaction feedback should be used to establish a baseline against which to measure improvement. Future feedback should be used to ensure continuous improvement.
4. After the initial customer-satisfaction survey, surveys should be conducted continually on a monthly, quarterly, or annual basis. The organization should conduct surveys often enough to be well informed, but not so often that the process becomes burdensome.
5. Customer delight is a level above customer satisfaction. A satisfied customer is happy with a product or service, but will still consider buying from the competition. A delighted customer won't even consider the competition.

KEY TERMS AND CONCEPTS

Baseline

Customer delight

Customer panels

Customer satisfaction

Customer-satisfaction data

Customer-satisfaction survey

External customers

Internal customers

Negative advertising force

Positive advertising force

REVIEW QUESTIONS

1. Explain the rationale for determining customer satisfaction.
2. Describe the most common approaches to determining customer satisfaction.
3. How should customer-satisfaction information be used?
4. When should follow-up surveys be conducted after the initial survey?
5. Explain the difference between customer satisfaction and customer delight.

ENDNOTES

1. Charles C. Poirer and William F. Houser, *Business Partnering for Continuous Improvement* (San Francisco: Barrett-Koefler Publishers, 1993), 213.
2. David R. Altany, "Bad Surveys Flood the Market," *Industry Week*, September 20, 1993, 12.
3. Philip E. Atkinson, *Creating Cultural Change: The Key to Successful Total Quality Management* (San Diego: Pfeiffer & Company, 1990), 119.

CASE STUDY 11–1

Determining Customer Satisfaction at MTC

Before arriving at a decision on the approach MTC would take in collecting customer feedback, John Lee discussed the issue at length with MTC's facilitator. The approach selected was the customer panel. The Steering Committee would develop a representative sample of MTC's customer base and invite individuals in the sample to become members of a customer panel. The facilitator was given responsibility for organizing the panel and collecting feedback.

The facilitator decided to develop a telephone survey instrument and train several graduate students to actually conduct the survey. Before beginning the survey, he pilot-tested the instrument himself on ten members of the customer panel. As a result of the test, he shortened the survey instrument and revised two of the questions.

When he was happy with the survey instrument, the facilitator trained his team of graduate students in the fundamentals of conducting telephone surveys. Data collection took a week, and compiling and summarizing the results took another week. The summary was turned over to John Lee and the Steering Committee for analysis.

By studying the summary, Lee and the Steering Committee members were able to identify several ways to improve their power supplies. Areas in need of improvement were as follows:

1. Better on-time delivery rate
2. Lighter-weight chassis
3. Longer product life
4. Smaller size

Lee and the Steering Committee members now had information with which they could work. Each area identified could be improved. Lee and the committee felt comfortable MTC's personnel could make the needed improvements. Working with the Steering Committee, Lee decided to take the following action:

1. Write a personal letter to all customers notifying them of the improvements MTC was making to its power supplies and thanking them for their business.
2. Dispatch MTC representatives for face-to-face follow-up conferences with the company's largest customers. The message to be communicated was: "MTC appreciates your business and wants to make sure you are completely satisfied."
3. Later in the implementation, assign the various areas needing improvement to cross-functional teams.
4. Establish a date for completion of improvements and conduct a follow-up survey.

In the meantime, Lee and the Steering Committee decided to measure internal customer satisfaction. The facilitator convened an internal customer panel representing all of MTC's departments. With their help he developed a survey instrument that was distributed to all employees. The facilitator then used the internal-customer panel as a sounding board for analyzing employee feedback.

The summary of the internal survey revealed the following areas in need of improvement:

1. More opportunities are needed for employee input into decisions about how to do work assignments better.
2. More training is needed for line personnel.
3. More and better communication/cooperation is needed between the Sales and Marketing departments.

John Lee and the members of the Steering Committee were pleased with the results of the survey. Clearly there were problems, but the problems could be dealt with and Lee was determined to begin immediately. His confidence in MTC's ability to survive and succeed was growing day by day.

CASE STUDY 11–2

Dissatisfied Customers at ESC

Lane Watkins knew there was no need to measure the satisfaction of ESC's internal customers. His recent survey of employee attitudes left little question as to what the results would be. However, a survey of external customers might work. He called a meeting of the Steering Committee to discuss the issue. A consensus quickly formed around the opinion that a survey was unnecessary.

Said one member, "At ESC, dissatisfied customers are the rule, not the exception." The other members nodded their heads in agreement. Watkins tried to move the group toward forming a customer panel, but got nowhere. The thinking of the Steering Committee was that it would require a lot of work and trouble, and the results were already well known. After the Committee discussed the issue for over an hour, Watkins could see that no progress would be made, so he adjourned the meeting. There would be no baseline survey of customer satisfaction.

Tailor the Implementation

- Rationale for Planning the Implementation
- Planning the Implementation as a Steering Committee Function
- Reviewing Resources, Both Physical and Intellectual
- Selecting the Best Approach
- Incorporating the Plan/Do/Check/Adjust (PDCA) Cycle
- Planning as an Ongoing Process
- Management of the Total Quality Process and Activities

In this step the implementation is tailored to the organization in question. Little has been written about implementing Total Quality due to the prevailing philosophy that, since all organizations differ in important ways, no two organizations can implement Total Quality in the same way. While this is true, our philosophy is that a common set of actions needs to be undertaken in a specific order, regardless of individual differences in organizations. Many of the actions recommended in this chapter are appropriate for any organization. The section entitled "Selecting Your Approach" shows how organizational differences can be accommodated through tailoring the implementation.

We must also point out that this is the point at which the implementation is most likely to break down. Obviously, the approach taken must fit the organization's needs, culture, strengths, and weaknesses, but *fit* is not the only potential problem. Failure to plan properly once you select the best approach is guaranteed to cause problems. This is an area that requires the best work of the Steering Committee.

RATIONALE FOR PLANNING THE IMPLEMENTATION

Successfully implementing Total Quality is one of the most difficult initiatives an organization can undertake. So many factors mitigate against success that an organization simply cannot get by without effective planning. Some of the roadblocks an organization

Figure 12–1
Roadblocks to Implementing
Total Quality

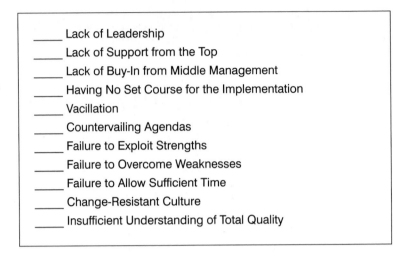

_____ Lack of Leadership

_____ Lack of Support from the Top

_____ Lack of Buy-In from Middle Management

_____ Having No Set Course for the Implementation

_____ Vacillation

_____ Countervailing Agendas

_____ Failure to Exploit Strengths

_____ Failure to Overcome Weaknesses

_____ Failure to Allow Sufficient Time

_____ Change-Resistant Culture

_____ Insufficient Understanding of Total Quality

is likely to face in trying to implement Total Quality are listed in Figure 12–1 and discussed in the following paragraphs.

Lack of leadership and lack of support from the top will sidetrack the Total Quality implementation before it gets started. The organization would not have gotten to this step without leadership and support from the top. If there has been progress up to this point, the support will probably continue. However, if Steps 1 through 11 have not been handled well, support may dry up. When management sees no positive result, its resolve gives way to doubt. Doubt, in turn, can result in the abandonment of Total Quality. This scenario can be avoided by selecting the appropriate implementation scheme and planning thoroughly. The implementation of Total Quality cannot be left to trial and error, guesswork, or intuition. Decisions and plans must be made on the basis of facts.

Lack of buy-in from middle management can also be a serious problem. Middle managers must be considered when the Steering Committee does its planning. Middle managers are the members of the team who are likely to feel most vulnerable as the organization moves toward Total Quality, so they must be made to feel that they are still important and needed. The Steering Committee should share the plan with them and make them part it.

A Total Quality initiative that has *no set course for implementation* is no initiative at all. What is required at this point is thorough and thoughtful planning, based on coherent instructions from the Steering Committee. No progress will be made if the effort goes in several directions at once. This issue comes back to leadership. The organization must speak with a single voice.

Vacillation can lead to poor or inconsistent results and should be avoided. Again, this is closely linked to planning and leadership. The Steering Committee must plan where the organization is going and demonstrate leadership in consistently following the plan. Note, however, that if a plan clearly isn't working, the Steering Committee should call it off. It must then regroup, examine what went wrong, and, using that data, come up with a new plan. This often happens at halftime in football games; if the game plan isn't working, the coaching staff steps in and makes revisions.

Countervailing agendas among managers of high rank, even members of the Steering Committee, are not uncommon, yet can be powerful deterrents to the Total Quality implementation. When such managers are part of the team, the senior manager must take whatever action is necessary to either bring such individuals around or remove them from the team.

Failure to exploit strengths and overcome weaknesses is a common mistake made by organizations implementing Total Quality. The identification of organizational strengths and weaknesses was done in Step 8. Successful implementation requires that the Steering Committee make use of this information by planning how to exploit the organization's strengths and accommodate its weaknesses.

Failure to allow sufficient time for the benefits of Total Quality to develop is a major deterrent to successful implementation of Total Quality. Patience is not often a virtue of Western organizations. This is why so many operate with such short-term horizons. Managers tend to be more interested in results for the next quarter than for next year or the next five years. Hence, it is not surprising that many organizations have difficulty developing the long-term perspective required to successfully implement Total Quality. Inherent in the decision to proceed with Total Quality must be rock-solid assurance that adequate time will be allotted for changes to succeed. Remember, although some improvements can be achieved quickly, the real benefits of Total Quality typically are slow in coming.

An organizational culture that is resistant to change is also to be expected. We believe that cultures are universally resistant to change. An organization's culture typically evolves over long periods of time and is influenced by leadership, personalities, objectives, and the environment in which the organization exists. Consequently, it also takes a long time to change. Most employees have come to expect that when change is introduced in an organization, some individuals and groups benefit while others lose. Maintaining the status quo often appears preferable to risking coming out a loser, so many people try to prevent change from taking place. In the case of change brought about by Total Quality, however, there should be no losers.

Although by now the Steering Committee has been thoroughly trained, the rest of the organization may have an *insufficient understanding of Total Quality*. Implementation planning should include activities to ensure that employees understand the *what* and *why* of Total Quality and how they will be affected by the changes.

Planners must take into account all of the known potential inhibitors and devise means of dealing with them. A well-planned implementation takes a structured, orderly approach in which issues are anticipated and dealt with in advance. A poorly planned implementation will be sidetracked by every issue that comes up, and will make little progress.

PLANNING AS A STEERING COMMITTEE FUNCTION

Having completed the first 11 steps, the Steering Committee is well experienced by now and has much of the information needed to make a rational judgment about how the implementation should be tailored.

A question that comes up frequently when we talk to groups about Total Quality is, "Isn't management giving up its responsibility when it gets involved with employee empowerment and involvement?" The answer is an unqualified NO. Total Quality involves empowering all employees to take control of their jobs getting them involved in decisions about improvements to their job processes. Note that every action taken up to this point has been the result of Steering Committee action. In fact, it is not until Step 16 that empowered/involved employees begin to take over at the team level. Step 12, tailoring the implementation, is one of the most important management functions in the entire implementation process, because the future of the organization can depend on it. In this step, the organization is being managed by those paid to manage. Ultimately, once Total Quality has been successfully implemented, employee empowerment and involvement means that high-level managers are better able to focus on high-level decision making.

REVIEWING RESOURCES, BOTH PHYSICAL AND INTELLECTUAL

In the preceding four steps, the Steering Committee listed the organization's strengths and weaknesses, identified who in the organization can be expected to become advocates of change and who will be resisters, and developed a thorough knowledge of employee and customer satisfaction. As its first step in tailoring the implementation, the Steering Committee should review the information collected thus far. The holistic view of the organization that results will help determine how best to approach the implementation. Should the organization begin the implementation through statistical process control (SPC), production methods, or business processes? Should a specific department be selected to begin the initiative?

The process of planning Total Quality activities that exploit an organization's strengths while accommodating its weaknesses will lead the organization toward one implementation approach and away from another. Similarly, knowledge of which employees are likely to be supportive can suggest using one group while avoiding another, at least in the initial efforts. Customer satisfaction comes into play in suggesting specific areas of improvement on which to focus initial efforts.

In addition to considering the intellectual resources discussed so far, it is important to consider physical resources. For example, if the organization has a system to track and analyze product defects, improvements in that system might be included as part of the initial effort. Note, however, that it is usually inadvisable to purchase new equipment in the initial phase of implementation. It is better to use the equipment on hand until the Total Quality effort is well established. Few things will sink a Total Quality initiative faster than having to spend a lot of money on it up front. Fortunately, up-front capital expenditures are rarely needed.

SELECTING THE BEST APPROACH

There is no magic formula for successfully implementing Total Quality. Since organizations differ so much, a one-size-fits-all implementation plan would fit no organization at all. Nevertheless, ideas that apply to your situation can be found in the experiences of other organizations. It is important to learn as much as you can about what has worked and what has not. You will find quality-related trade journals to be invaluable resources for ideas. Try to relate the experiences of others to your organization. Keep in mind that you are not looking for the "right" approach. Many different approaches have merit and can be tailored to work for your organization.

Following are some guidelines to follow in selecting an approach and tailoring it to your own organization.

Exploit Your Organization's Strengths

An approach that makes best use of the organization's strengths and involves those groups or individuals who are most likely to be enthusiastic supporters will have the best chance of success. It is up to the Steering Committee to evaluate various potential approaches to decide which is likely to work best for the organization. Note that although this is a Steering Committee responsibility, outside help from a consultant is not ruled out. In fact, it may be appropriate, but keep in mind that *you* are going to have to live with any decision made. Consequently, all decisions are ultimately the responsibility of the Steering Committee.

As an example, the Steering Committee of one firm that was strong in data collection and analysis decided on a unique approach that worked well for them. They began the implementation by identifying all their key processes and corresponding sub-processes. Their list was as follows:

Engineering Processes: Hardware design, software design, design for manufacture, engineering change.

Production Processes: Electronic assembly, mechanical assembly, software integration, testing.

Business Development Processes: Customer liaison, opportunity discovery and tracking, negotiation, program initiation.

Business Processes: Financial/accounting, human resources.

Support Processes: Management Information System, facilities.

This list is by no means complete and is provided only as an example. There were more key processes and sub-processes than those listed, and each entry had several sub-sub-processes.

Next, the organization formed a high-level cross-functional team for each process. They required each team to develop a flow chart of its key process and begin improving the process and all of its corresponding sub-processes. As this process-team activity got underway, each of the teams formed sub-process teams to take on specific tasks. Since at the outset all of the organization's processes were included, very quickly the entire company was involved in the implementation of Total Quality. Today this organization is a world-class company.

Could you use this approach? If your organization has expertise in data collection and analysis, perhaps so. Why are data collection and analysis critical to this approach?

Because you must be able to collect and analyze data related to the processes in question to establish a benchmark for the process and to determine which action might affect which parameter. Then you must be able to repeat these steps over and over as changes are made to be sure the changes are taking you in the right direction. If data collection and analysis are organizational weaknesses, you should select a different approach. However, you will need to start working on this shortcoming immediately, because you will need the capability later.

We should point out here that the company that is the subject of our example took on an enormous challenge by including all of their processes in the initial implementation. For many organizations, it may be preferable to start with one process group and expand into others only after the initial is operating successfully.

If you have a department that shows an interest in Total Quality and its key employees are probable supporters, you might tailor the initial implementation activities to that department's functions. For example, Just-in-Time (JIT) production techniques might be implemented in manufacturing. This activity flows easily into work teams for the JIT cells, and these teams become problem-solving units for their processes. Once experience is gained, it will be relatively easy to export successful techniques to other departments.

The most popular approach for kicking off the implementation is the forming of cross-functional teams to find solutions to specific problems. In the early days of the Total Quality movement, these teams were often formed by employees without the direction of management. We have come to the conclusion that such teams should be formed by the Steering Committee, so that it can ensure that the teams attack problems in accordance with the vision and goals of the organization. After the organization gains experience with Total Quality, it may be possible to reduce the level of control by the Steering Committee, but initially it is important that the Steering Committee keep implementation efforts focused. This is not to say that the Steering Committee members should not be receptive to suggestions from within the organization. This is essential. Employees must believe that the Steering Committee will listen to their suggestions and act on them. The issue here is that high-level guidance needs to be provided during the initial stages of implementation.

The customer-satisfaction method is another approach. Remember that the term *customers* refers to both internal and external customers. Implementing Total Quality from the perspective of internal customers is relatively simple. To begin this approach, you need only survey your internal customers. Using external customers as the basis of an implementation approach is more difficult. The logistics may make such an approach impractical. Since the feedback loop from the external customer is much longer than that of the internal customer, it is more difficult to measure the response to changes as they are made. With internal customers, responses to attempted improvements can be measured immediately. Teams can be deployed to find ways to improve processes, and feedback on the reactions of internal customers can be collected quickly. Experience has shown that an organization that improves internal-customer satisfaction also improves external-customer satisfaction.

The list of possible approaches is long. The examples explained herein are just that: examples. It is critical that each organization adopt an approach that is tailored to its specific strengths and weaknesses.

Accommodate Your Organization's Weaknesses

Every organization has weaknesses that will have to be accommodated in the early stages of implementation. For example, if an organization's middle managers are all from the "old school" and cannot bring themselves to believe that their subordinates have the ability to think, then employee involvement efforts are going to be difficult. In such a case it will be best to avoid the issue for the moment and find another implementation approach. However, you will want to begin working to change your middle managers' viewpoint, because otherwise they will eventually become an impediment to full implementation. This is just one example of a weakness an organization might need to accommodate during the initial stages of implementation. Regardless of what the actual weakness is, it should be handled as illustrated by this example.

Consider Your Organization's Culture

In addition to weaknesses and strengths, the overall culture of your organization should be considered when selecting an approach for implementing Total Quality. What kind of approach is most likely to "turn on" everyone in the organization? If you can find an approach that gets people excited, even those who are not directly involved, progress will come more easily. On the other hand, selecting an approach for which employees have little enthusiasm is apt to cast a shadow on the entire effort. Certainly much of the potential for excitement rests in management's leadership and its ability to help everyone understand what Total Quality will do for them.

Avoid Numbers Games

Approaches that involve "numbers games" should be avoided by organizations implementing Total Quality. A few years ago, the measure of merit in an organization's Total Quality program was how many teams had been formed and how many employees had been trained. These notions were misguided at best, and often doomed the implementation to failure. In such cases there was no assurance that team activities supported the organization's vision, if indeed the organization had a vision. Training only led to frustration because newly trained employees had no means to put their new knowledge to use. It was not just small, unsophisticated organizations that fell into this trap. Executives of a major electric utility company told us that the biggest mistake they had made in implementing Total Quality was training all 17,000 of their employees at the outset, when no more than a handful of them would actually be involved in Total Quality activities for at least two years. Avoid this mistake. It is far better to begin with one team, properly trained and working on a meaningful task that supports the vision, than with a thousand people put on teams that have no meaningful assignments.

Start small and build on your experience. Before long, you will be able to form teams for all the problems that can be identified. As experience is gained, teams can be added and given meaningful direction. Provide training on a just-in-time basis. If a team is to start next week, train them this week and no sooner. Form teams when needed to take on specific problems as directed by the Steering Committee.

Don't Delegate the Implementation

Another mistake frequently made by top management is delegating responsibility for the implementation to a specific individual or department. We are not aware of a single successful delegated implementation. We observed one such an attempt over a period of a few years. In this case, top management delegated the responsibility for implementing Total Quality to a nationally known expert who managed their Quality Assurance Department. These executives saw Total Quality as just another program to be endured. Since the issue was quality, it seemed logical to delegate responsibility to the expert in the Quality Assurance Department. We watched this person work upstream for almost three years in an attempt to implement Total Quality. Because everyone in the company knew that top management was neither committed nor involved, the inevitable roadblocks occurred and implementation failed. The nature of Total Quality is such that it must have active, unwavering support from the top. This means the top management and the Steering Committee.

Select for Success

Success is essential in the early stages of the implementation. Early failures will take away the steam needed to break cultural inertia. Consequently, we recommend that you select initial projects for which success is almost guaranteed. There are two reasons for this. First, getting the projects planned and put into operation and the participants trained and pulling in the same direction is difficult at best. Once teams successfully complete a few projects, they will learn how to function the Total Quality way. Second, early successes give the organization a psychological boost. On the other hand, if you start with tough assignments and fail, the Total Quality concept will lose credibility and the natural reluctance of people to change will be reinforced. Our advice is to pick initial projects that are meaningful yet offer a reasonable expectation of success. Gain some experience, then tackle the tougher projects.

INCORPORATING THE PLAN/DO/CHECK/ADJUST (PDCA) CYCLE

Once the implementation is started, the PDCA cycle comes into play. The PDCA cycle, illustrated in Figure 12–2, was developed by Walter Shewhart, Dr. W. Edwards Deming's mentor at Bell Labs. In his early lectures to engineers and managers in Japan, Deming stressed the use of the PDCA cycle as the logical means of continuous improvement. Although in Deming's model PDCA stands for Plan/Do/Check/Act, we, like many others, prefer Plan/Do/Check/Adjust, since *adjust* better describes what the final step is really about.

PDCA can be used with any product- or process-improvement project. Following the PDCA cycle, you *plan* the implementation, *do* the work, *check* progress, and *adjust* as necessary. Repeat the cycle continually throughout the life of the implementation and individual implementation projects. Continue to make adjustments that will improve the process. The need for PDCA never ends.

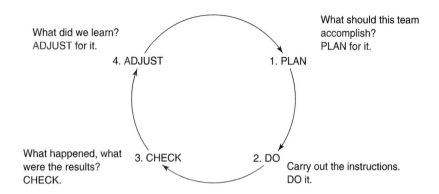

5. Repeat Step 1, having adjusted for the knowledge gained.

6. Repeat Step 2 with new instructions.

Figure 12–2
The PDCA Cycle

PLANNING AS AN ONGOING PROCESS

There will never be a time when the Steering Committee can say to itself, "We have done the planning. Now we can move on to something else." Total Quality is a journey that never ends. This may sound discouraging at first, but look closer. What it really means is that in our imperfect world, no organization can achieve peak performance all of the time, but it can be achieved consistently through a continuous, never-ending effort. Once the initial phases of implementation have been completed, organizations should go to the next level, expand into other areas, and respond to new challenges. Consequently, the need for planning by the Steering Committee never ends. What is interesting about this is that it takes us back to the traditional definition of management, which has planning as a major function. Somewhere along the line, management decreased its emphasis on planning and increased its emphasis on controlling. As a result, many organizations over-control while drifting aimlessly toward failure. This concept is known as *managing the organization to death*. Planning is the function that keeps organizations on the right track.

MANAGEMENT OF THE TOTAL QUALITY PROCESS AND ACTIVITIES

In this chapter we have presented the model for all Total Quality activities, and in fact for the entire organization. The model may be difficult to grasp at first, but it's important that it be fully understood. Once you implement the Total Quality model, old management techniques must be discarded. In the past, the business model was typically a CEO assisted by a staff of executives, often with their own individual departments, who managed the organization through individual, often countervailing, efforts. In the Total

Quality organization, the top managers, even though they may be the same people, serve as the Steering Committee and manage as a team. Each action must pass the test of whether it benefits the overall quest to achieve the organization's vision. The Steering Committee *plans* an action, assigns it to a team that is appropriately equipped and trained, then prepares to receive the team's feedback. The team *does* the planned action. The team's feedback is then *checked* to make sure things are going as intended, and if not the plan is *adjusted* to make it work better. This cycle repeats over and over for each planned activity and for newly planned activities. This is the pattern that will be followed as long as Total Quality is the philosophy by which the organization operates.

SUMMARY

1. Because there are so many potential roadblocks that can mitigate against success, it is important to plan the Total Quality implementation carefully and tailor it to the individual organization.
2. Responsibility for the implementation cannot be delegated. While involving and empowering employees is essential to a successful implementation, ultimate responsibility for the implementation rests with the Steering Committee.
3. When selecting the approach to be used in implementing Total Quality, it is important to choose one that exploits the organization's strengths, accommodates the organization's weaknesses, considers the organizational culture, avoids numbers games, and has the best chances of success.
4. Every implementation activity, including all team projects should incorporate the Plan/Do/Check/Adjust cycle.
5. In a Total Quality organization, planning never stops. Planning is what gives the organization direction as it moves toward achieving its vision.

KEY TERMS AND CONCEPTS

Change-resistant culture	Lack of leadership
Countervailing agendas	Lack of support
Failure to allow sufficient time	No set course
Failure to exploit strengths	Numbers games
Failure to overcome weaknesses	Plan/Do/Check/Adjust cycle
Insufficient understanding of Total Quality	Vacillation
Lack of buy-in from middle management	

REVIEW QUESTIONS

1. Explain why planning is so important to a successful implementation.
2. Why is planning the implementation a Steering Committee function?

3. Describe how you would go about selecting the best approach for implementing Total Quality in an organization.
4. Explain the Plan/Do/Check/Adjust cycle and how it is used for implementing Total Quality.
5. Why is ongoing planning necessary in a Total Quality organization?

======= **CASE STUDY 12–1** =======

Tailoring the Implementation at MTC

The time had finally arrived when the implementation of Total Quality could take a major step forward. John Lee scanned the faces of the Steering Committee members seated around the conference table. They were as eager as Lee to take the next step. These top managers knew MTC's strengths and weaknesses. Today's meeting had been called for the purpose of selecting an implementation approach that was tailored to MTC's specific characteristics.

Lee began the meeting by distributing a handout that contained the following criteria:

1. Exploit strengths.
2. Accommodate weaknesses.
3. Consider the organizational culture.
4. Avoid numbers games.
5. Select for success.

After discussing several different approaches, the Steering Committee seemed to be forming a consensus around establishing a cross-functional team to take on the task of reducing scrap and waste in the metal forming and machining unit. The employees in this unit had already identified this as a problem that needed to be confronted, and had already had made some excellent suggestions for improvements. Lee agreed that this was probably the place to start, but he had one concern. Howard Lemon, a senior machinist, had been identified as a potential inhibitor. Howard was a good machinist, but he had a history of resisting change. The battle to convince him to learn computer-numerical control machining, although eventually won, had been difficult.

After discussing various options for over an hour, the Steering Committee found a way to turn their Lemon into lemonade. For some time the head of the Manufacturing Department had been considering starting a training program in machining. He simply could not find well-trained entry-level machinists and had reached the conclusion that MTC was going to have to train its own. Why not put Howard Lemon in charge of developing the program? This solution would kill two birds with one stone. Other Steering Committee members agreed and the decision was made.

Discussion quickly focused on naming members of the cross-functional team and giving the team its chance. John Lee could sense the excitement in the room. Real progress was being made.

=== **CASE STUDY 12-2** ===

Planning Problems at ESC

Lane Watkins didn't know what to expect when he called the Steering Committee meeting to discuss developing a tailored implementation plan for ESC. In fact, he had had doubts as to whether anyone would even show up. Although some members didn't attend, Watkins felt he had enough members present to proceed.

Watkins asked Steering Committee members to brainstorm ideas for an initial project that would exploit ESC's strengths, accommodate its weaknesses, and have a high probability of success. For about 45 minutes there was enough discussion going on that Watkins began to think they might actually get somewhere. But his hopes were quickly dashed. One member of the Steering Committee who had been silent throughout the meeting said, "It doesn't matter what approach we select, it isn't going to work. So why don't we stop wasting our time?" He then got up and left the room, and all the others, except Lane Watkins, left with him.

Identify Projects

At first glance, this step might appear to be a rehash of Step 12, because several of the main topics resemble those of the preceding steps. But this step is more narrowly focused on specific projects rather than on the macro view necessary to select an approach to the overall Total Quality implementation. Having selected our approach, we now find ourselves at the point of choosing specific projects for making organizational improvements. The tests applied to potential projects are essentially the same as those applied when selecting the implementation approach. The key is to select projects for which you have the necessary expertise and willing workers who are likely to be favorably disposed to Total Quality. All projects must be supportive of the organization's vision and broad objectives, and, at least with the early projects, must have a high probability of success.

RATIONALE FOR SELECTING PROJECTS

The question is not whether projects should be selected; they must be. The question is *who* should select them, and according to what criteria. In the early 1980s, many Western companies adopted Quality Circles, typically without first understanding the big picture of Total Quality. Western industrialists who toured Japanese factories came back with stories of how Quality Circles were responsible for the dramatic gains made by

---

TOTAL QUALITY TIP

Organizations Are Made Up of Teams

"Quality circles often fail because they are not appropriately focused; their accomplishments in these domains are not integrated with strategy. . . . The circles made a few workplace improvements and changed some operational methods. However, management rejected many of their proposals as either 'too costly,' or 'unnecessary.' Reports were ignored, requests for funding denied, and many attempts to make local changes were squashed."[1]

Japanese companies in quality and productivity. Although there were some notable exceptions, for the most part, Quality Circles accomplished little in the United States. When Quality Circles were set up, circle members were told that in addition to their normal jobs, they were to develop improvement projects. Usually, but not always, improvement projects had to be approved by management. Once a project was approved, it became the responsibility of the Quality Circle. For the most part, the circles had access to outside expertise when it was needed.

Typically, Quality Circles were given an hour or so each week to work on their projects. Except for that hour, each member continued to do his or her regular job. Worker acceptance varied, but in general, employees seemed to enjoy their involvement.

Where many Western Quality Circles failed, we believe, was in the areas of top-level leadership and commitment. The lack of direction given the circles, coupled with management's unwillingness to adopt the changes the circles recommended, inevitably led to failure. There was a general hue and cry from management that Quality Circles had been given the freedom to develop improvements, but all they were coming up with were such mundane projects as assigning reserved parking places, solving plumbing problems in the rest rooms, and so on. The astute reader will recognize this problem not as a shortcoming of the Quality Circles, but of management. This is why we believe that rather than having projects develop from the bottom up, it is better to assign them from the top down. This is not to say that lower-level employees should not suggest projects for management's consideration; they definitely should. But ultimately it is management's responsibility to determine which projects support the vision and objectives and which have the greatest potential benefit to the organization.

The issue of management's unwillingness to adopt proposed changes is another management problem. Without proper direction, Quality Circles pursued their own agendas. Then, when management did not approve of a change they proposed, such as the elimination of reserved parking spaces, the popularity and credibility of Quality Circles soon waned.

Don't get the impression that Quality Circles work in Japan but don't work in the United States. It is a fairly common practice in this country to misuse a new initiative and then claim it doesn't work. This is what happened with Quality Circles. Very few Quality Circles in the United States were like those in Japan.

Taiichi Ohno, the genius who created the Toyota Production System, also created what came to be known as Quality Circles. He grouped workers into teams, led not by a foreman, but by a team leader. All team members worked together in one part of the assembly process. The team leader coordinated the work of the team in addition to doing actual assembly work and filling in for absent team members—not a common practice in U.S. industry at the time. Ohno instructed the team that they were to determine how best to do their part of the production process—again, not a common practice in the United States. He also made the team responsible for housekeeping in their areas, for minor tool repair and maintenance, and for quality checking. Once the teams were functioning smoothly, Ohno set aside time for them to periodically discuss and develop ways to improve their processes.[2]

In the early 1980s there were no similar teams in North America. Too few exist even today. Notice that these teams were made up of people who worked closely together. Quality Circles in the United States were staffed by volunteers who seldom worked together in the same discipline or in the same physical space. It is important to note that Ohno gave his teams explicit instructions as to what their responsibilities were, and saw to it that those responsibilities were always associated with team members' own processes rather than being totally open-ended. Only after groups demonstrated the ability to function as teams did Ohno set aside time to discuss and devise ways to improve processes. In the United States, on the other hand, industry considered open-ended improvement as the *raison d'etre* for Quality Circles. Many firms in the United States completely misinterpreted the substance, purpose, and scope of the Japanese model.

All of this serves to demonstrate why the recommended approach is to have management select, with employee input, the improvement projects. As illustrated in Figure 13–1, without guidance from management, employees are not likely to pursue activities that contribute to the organization's vision and objectives. Teams at the working level are simply not in a position to decide what is best for the overall organization. This is still management's responsibility, and it always will be.

PROJECT SELECTION AS A STEERING COMMITTEE FUNCTION

In the previous section we spoke of management's responsibility to identify the projects that best support the organizational vision and objectives. By "management" we mean the Steering Committee, the CEO, and those who report directly to the CEO. With Total Quality, the top management team is responsible for steering the organization toward its vision by ensuring that every activity of the organization moves it in the right direction.

CONSIDERATIONS WHEN SELECTING PROJECTS

As the Steering Committee considers potential projects, several criteria should be applied. These are illustrated in Figure 13–2 and discussed in the following paragraphs. The criteria are similar to those applied in the previous step, when the Steering Committee selected the best approach for the implementation.

Figure 13–1
Employees Need Manage-
ment Guidance in Pursuing
the Organization's Vision

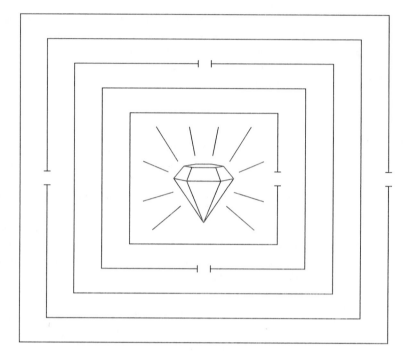

The Organization's Strengths and Weaknesses

The initial projects must be compatible with the strengths of the organization. Select projects you can support with expertise, and avoid those that might be adversely affected by organizational weaknesses.

The Personalities of Those Involved

Every task must be carried out by people. As discussed in the previous step, some people will be supportive of Total Quality and some won't, at least not at first. One of the main considerations in the selection of initial projects is the availability of employees with the expertise and *positive attitudes* necessary for success. Do not make the mistake of selecting projects for which there are not likely to be enthusiastic team members. Expertise will not overcome an unwillingness to perform. If it comes to this, choose enthusiasm over expertise. Of course, it is best to select projects that can be assigned to experts who are also enthusiastic supporters.

The Organization's Vision and Objectives

A primary function of the Steering Committee is to ensure that all projects are supportive of the organizational vision and objectives. There will always be more project opportunities than organizational resources. Consequently, it is imperative that those that have the greatest value be given priority. The question that must be asked is, "Does this proposed project support the vision and broad objectives?" If the answer is no, the project should be dropped. If it cannot pass the test of supporting the vision and objectives, then it is of no value to the organization, and may even be counterproductive. Be firm on this point.

The Project's Probability of Success

In Step 12 we pointed out that the approach selected for the implementation must offer the greatest chance for success. The same is true when selecting the early improvement projects. This point cannot be overstated. A few early successes will overcome inertia and provide the momentum needed for success. Correspondingly, failures in the initial projects can galvanize resistance and thereby ensure failure of future projects. Consequently, it is wise to select projects during the early stages that offer near certainty of success, considering the organization's strengths and weaknesses, employee attitudes, and the project's degree of difficulty. Don't start with "home run" type projects. It is far better to plan for continuous, incremental improvements.

SOLICITING SUGGESTIONS FROM ALL SOURCES

Since your organization is just getting started in Total Quality, your employees may not offer many suggestions at first. This is normal. Employee reluctance to offer suggestions

Figure 13–2
Considerations in Project
Selection

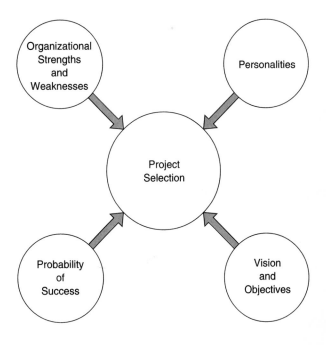

grows out of being conditioned to the old style of management in which managers (bosses) directed all aspects of the workers' functions and believed that workers should work and managers should think. This style is commonly known as Type X management. It is captured in the familiar phrase, "Don't make waves, just follow orders."

Japanese organizations suffered from the same conditioning before they adopted the Total Quality philosophy. Today, Toyota is recognized as a world leader in Total Quality, but in 1960 Toyota received only about one suggestion per employee per year. As Total Quality took hold at Toyota, this changed remarkably. In 1982, Toyota received almost 33 suggestions per employee.[3] Equally important, 95 percent of the employees' suggestions were implemented, compared to only 33 percent in 1960. The typical Western organization still receives fewer than one suggestion per employee per year.

One of the cornerstones of Total Quality is the recognition that *all* employees are capable of proposing valuable ideas. We are beginning to understand that workers who are faced with the same problems day in and day out are well equipped to help solve those problems. Why don't they just do it, then? There are several reasons, ranging from employees' lack of incentive to make improvements to their not being allowed to make improvements. This situation changes with Total Quality. With Total Quality, management solicits the ideas of all employees, considers each one seriously, and gives feedback to employees who make suggestions.

There are two problems facing managers who seek ideas from all employees. The first is overcoming the ingrained culture on both sides, wherein managers don't think employee's ideas are worthy of consideration, and employees have learned to keep their ideas to themselves. The second is sustaining the idea flow once it starts. This is done by responding in a timely manner to every suggestion, telling the submitter that it will be implemented, or that it needs to be modified and then implemented, or that it has been seriously considered but cannot be implemented, and why. Responding in a timely manner can be far more difficult than one might assume. We are told by employees of many companies we visit that responses to employee suggestions (one way or another) are not timely, even in companies that are well along in implementing Total Quality. Without a timely response, employees will assume that management isn't interested in their ideas, and the flow of suggestions will stop.

As the Steering Committee is selecting the initial projects, consideration should be given to ideas that flow up from employees. The Steering Committee should give as much attention to the assembler's suggestion as to that of the CEO. Just apply the same tests. Does it support the vision? Does it exploit an organizational strength? Will the employees involved pursue it enthusiastically?

DEFINING THE PROJECT'S CONTENT, SCOPE, AND TERM

Once the Steering Committee has found a likely candidate for a project, the next step is to define it. The project must have a clearly stated objective, one that does not leave room for interpretation by team members or even by the Steering Committee. There must be clear beginning and ending points. The scope of the project, particularly early ones, should be such that they can be completed within a few weeks or less. The Steering Committee should provide a suggested schedule for the project, but must also recognize that rigorous adherence to the schedule while everyone is still in a learning phase may not be possible. The project must be clearly defined in terms of its beginning, end, and content so that all parties understand what is expected.

SLOW AND STEADY WINS THE RACE

It is better to start out slowly and deliberately than to take on more than you can handle. Don't allow yourself to become enamored of the *number* of teams working on projects. Rather, pay strict attention to the progress of a few teams until all of the people involved become knowledgeable of the process. After employees have demonstrated their ability to work in teams, and the Steering Committee has demonstrated its ability to manage the process, you can expand the effort. However, to do so too suddenly can be chaotic and cause the process to fail. Proceeding slowly and surely is the best approach in the early stages of implementation.

SUMMARY

1. The rationale for selecting early projects carefully is to promote a successful implementation. The question is not whether projects should be selected, but by whom and according to what criteria. The Steering Committee is responsible for selecting projects.
2. In selecting projects the Steering Committee should consider the following: the organization's strengths and weaknesses, the personalities of those involved, the organization's vision and objectives, and the project's probability of success.
3. Although the Steering Committee is responsible for selecting projects, it is important that members solicit input from employees. Once given, employee suggestions should be acknowledged promptly and acted upon in some way.
4. It is important to clearly define projects so that all stakeholders know what is expected. Projects must have a definite goal, a definite beginning, and a definite end.
5. It is best to start implementation slowly. Avoid the temptation to try to hit a home run your first time at bat. Continuous, incremental improvements will bring better long-term results.

KEY TERMS AND CONCEPTS

Defining the project

Employee suggestions

Organization's strengths and weaknesses

Personalities of those involved

Probability of success

Quality Circles

Start slowly

REVIEW QUESTIONS

1. Explain why the Steering Committee must select projects.
2. Describe the impact of organizational strengths and weaknesses on project selection.
3. How do individual personalities figure in when selecting projects?
4. Why is it important to solicit suggestions for projects from employees?
5. What is meant by "the project should be clearly defined?"
6. What is the rationale for starting implementation slowly?

ENDNOTES

1. Greg Bounds, Lyle Yorks, Mel Adams, Gipsie Ranney, *Beyond Total Quality Management* (New York: McGraw-Hill Series in Management, 1994), 160.
2. James Womack, Daniel T. Jones, and Daniel Roos, *The Machine That Changed the World* (New York: Harper Collins, 1990), 56.
3. Tom Peters, *Thriving on Chaos* (New York: Harper Perennial, 1991), 88.

CASE STUDY 13–1

Selecting Projects at MTC

John Lee called the meeting so that Steering Committee members could select MTC's initial projects. Actually they would be discussing the *next* project. The first project—reducing metal fabrication waste—had been selected and defined, and the team that would work on the project had been named.

Now, Lee was anxious to select a project that had been suggested by employees. He wanted to promote employee buy-in and thought that implementing one of their suggestions would serve this purpose. After it had been decided that the first project would go to the metal fabrication unit, Lee had asked the other Steering Committee members to begin soliciting input from employees company-wide. Some of the suggestions offered would do little to support MTC's vision, but many of them were top-notch. Regardless of the quality of their suggestions, all employees submitting a suggestion received prompt feedback.

Lee distributed a handout containing brief descriptions of the suggestions he felt were the most worthy, and the Steering Committee members began discussing them. It didn't take long to reach consensus. Several employees had suggested trying to improve internal customer relations. Although employees at MTC generally got along well, frequently there was counterproductive friction among internal customers about the quality of work. More specifically, the chassis department and the circuit board department often butted heads over assembly issues.

The Steering Committee decided that improving internal customer relations between the chassis and circuit board departments would be a worthwhile project. The next step would be to decide who should be on the team.

CASE STUDY 13–2

Selection Problems at ESC

Lane Watkins was just about fed up with his assignment to implement Total Quality at ESC. Since there was no commitment from the top, nobody took the effort seriously— including the so-called Steering Committee. For Watkins, their last meeting had been the last straw. When the Steering Committee wouldn't even cooperate in selecting an implementation approach, he knew he had to either throw up his hands and quit or become more directive. Watkins chose the latter option.

He called a meeting of the Steering Committee to select ESC's initial project. When all members were present, Watkins began: "The time has come to select a project. If the implementation was being handled properly, we would select a project that supported ESC's vision and had a better-than-average chance of success. However ESC has no vision, and any project we select is likely to fail because there is no top-level commitment to this so-called implementation. Be that as it may, I was given this assignment and I am going to proceed as if there were a real commitment. If anyone has a problem with this, I suggest you make an appointment with the boss and tell him how you feel. And by the way, good luck! I've been trying to make an appointment with him ever since he came to me with this project, but so far he's been The Invisible Man."

Watkins continued: "The purpose of today's meeting is to select an initial project. Rather than waste any more of your time in pointless discussions that go nowhere, I have chosen the initial project myself: to improve internal customer relations among our major departments. Any questions?"

The group sat in stunned silence. Finally, Watkins said, "Have a nice day," and walked out of the room. The Steering Committee members followed shortly thereafter, in a mood of uniform disbelief.

Establish Team Composition

- Rationale for Team Composition
- Establishing Team Composition as a Steering Committee Function
- An Example of Establishing Team Composition
- Choosing Total Quality Advocates for Initial Teams
- Developing a Team Member Checklist

Now you have selected a project that is in concert with the organization's vision and broad objectives and has a high probability of success. The next step is to establish the team that will carry out the project.

RATIONALE FOR TEAM COMPOSITION

There are different kinds of teams. Only one kind of team entails automatic membership: the so-called *natural work team*. This is the kind of team Taiichi Ohno formed as the genesis of what came to be known in the United States as Quality Circles. Natural work teams consist of people who work closely together in the everyday performance of their jobs. Most Total Quality organizations will find these natural work teams useful, but not in the earliest stages of implementation. If your implementation approach makes use of natural work teams, these groups will be the teams that carry out improvement projects.

Most implementation approaches start by establishing *cross-functional teams*. Cross-functional teams are made of employees from a variety of departments or disciplines. The initial cross-functional teams lay the foundation that will help natural work teams carry out process improvements related to their functions. Once weaknesses have been identified by the cross-functional team, they can be eliminated by the appropriate natural work teams—the owners of the process.

Both types of teams have a place in a Total Quality organization. But whenever a project involves more than one department, the team composition must be cross-functional. Even if only one department is involved, a cross-functional team may be appropriate as a way to incorporate diverse points of view.

There is nothing automatic about the composition of a cross-functional team. Selecting the right people for these teams is as crucial as selecting the right project. The idea is to staff the team with employees who have the right mix of specialized skills from across the spectrum of functions directly or indirectly related to the project in question. Team members should be compatible and supportive of Total Quality.

ESTABLISHING TEAM COMPOSITION AS A STEERING COMMITTEE FUNCTION

In the early days of Total Quality in the United States, it was not unusual for teams to form spontaneously, staffed without input from management. This was certainly true of the early Quality Circle movement. Organizations soon learned that this was the wrong approach, and that management had to be involved to ensure that teams were working on high-priority projects that supported the organization's vision. Management also has to ensure that teams have the best mix of employees. The management entity responsible for implementing Total Quality is the Steering Committee; so it is the Steering Committee that should establish the composition of teams.

AN EXAMPLE OF ESTABLISHING TEAM COMPOSITION

The following example illustrates how the process of establishing team composition works.

Vigraph Corporation, a leading manufacturer of video projection systems, had a customer service problem associated with the shipping of parts and accessories. Customers complained that when they ordered a part, they frequently received the wrong item, or it was delivered later than promised, or it didn't work. Top managers at Vigraph knew that, if left unchecked, this situation would damage the company's reputation and ultimately its market share. Vigraph executives decided that their first Total Quality project would be to solve this problem. They decided that a team should be given the assignment of solving this problem. The team was asked to proceed as follows:

1. Create a flow diagram of the current process to show what happens from the point where a customer orders a part until the part is received by the customer.
2. Examine the flow diagram to determine where the process breaks down and how it can be improved.
3. Make recommendations to the Steering Committee for ensuring that customers get the right part, on time, every time, and that the part works.

The project itself ruled out the possibility of a natural work team, because multiple functional departments were involved in the process. At least three departments are involved in each transaction. Customer Service takes the order from the customer. The order goes to the Stockroom, where the part is withdrawn. The Testing Department may get involved if the part has not already been tested. Finally, the Shipping Department packs and ships the part. Clearly, a cross-functional team was needed. The team needed

TOTAL QUALITY TIP

Foolproofing Your Processes

"The Japanese word for it is 'poka-yoke.' It means 'foolproof mechanism.' Poka-yoke is used to help the operators work easily by preventing any errors from affecting their output to the next process or function. This prevents defects at the source. Originally applied to machines, poka-yoke can be applied to any process. Only a foolproofed process will completely eliminate operator mistakes that cause defects. Keep foolproofing, or poka-yoke, in mind as you analyze and improve your processes."[1]

Kiyoshi Suzaki

participation from Customer Service, the Stockroom, and the Shipping Department. It also seemed appropriate that the Accounting Department participate, since they became involved once the process was completed. In addition, a representative from the Testing Department could provide insight concerning faulty parts. With one member from each of these departments, the team would have five members. This left room for one or two more members without making the team too large. Six to eight members is ideal for most cross-functional teams. Vigraph's managers felt they should also bring an independent viewpoint to the effort. They considered membership from the Manufacturing, Sales, and Quality Assurance Departments. Their final decision was to have a representative from the Sales Department serve on the team. The result was a six-member cross-functional team made up of one member each from six departments: Customer Service, the Stockroom, Shipping, Accounting, Testing, and Sales.

Once the departments to be represented were identified, the next task was to select a person from each department to serve as a team member. The temptation here is to call on each respective department manager, but this isn't always the best approach. When a process is breaking down and allowing various kinds of mistakes to cause defects in customer shipments, the people most likely to know how to make the process foolproof are the ones who work inside the process every day. These are the order takers, stockroom clerks, and shipping clerks, not the managers.

Vigraph's decision was that the Customer Service Department would be represented by a hands-on clerk, as would the Shipping Department and the Stockroom. The Testing Department would be represented by a technician who routinely tested parts that go directly to customers. Accounting would be represented by an Accounts Receivable clerk, and a representative from the Sales Department would round out the team.

Deciding what *kinds* of people should represent the various departments on the cross-functional team was the easy part. The more difficult part was selecting the actual people. Choosing actual team members tends to be a subjective undertaking. A special effort should be made to be as objective as possible. Vigraph made a preliminary selection of individuals. However, since this was the company's first attempt at establishing a team, all potential team members were interviewed to get a feel for how well they would fit, and to determine their level of interest in participating. During one interview, a potential member claimed that Vigraph didn't need a team at all. His view was that if Vigraph would just "fire the dummy on the shipping dock and hire someone who could read English," the problem would be solved. It turned out that the individual in question was no dummy, and that he could read English quite well. Because this prospective member demonstrated such a negative, counterproductive attitude, he was not selected for the team. Instead, another member who worked in the same department but had a better outlook was selected.

Note that Vigraph used a cross-functional team to solve its customer service problem. No single department could have hoped to solve the problem because it crossed so many department boundaries. This will be the case with most problems. A cross-functional team can expose process weaknesses that a natural work team might not see or might not admit.

CHOOSING TOTAL QUALITY ADVOCATES FOR INITIAL TEAMS

Not everyone will be an advocate of Total Quality in the early stages of implementation. Some will fear that Total Quality may adversely affect their position or jeopardize their continued employment. Others will take a wait-and-see attitude. The Steering Committee should avoid both of these types of individuals when establishing early teams. There will be enough "growing pains" without the added problem of having disruptive or non-supportive members on teams.

DEVELOPING A TEAM MEMBER CHECKLIST

Figure 14-1 is a sample checklist for selection of team members. Your situation may demand additional attributes or may not require some of those listed. This checklist is intended as a starting point as the Steering Committee considers individuals for team membership.

It is important to keep in mind that a person who is not a likely candidate today may be right for another team later. This is especially true of people who adopt a wait-and-see attitude. However, once they see what Total Quality can do, they usually become supporters. For this reason, such an individual should not be blackballed forever from participation on a team. Consider the person again whenever his or her talents and skills seem appropriate. In fact, the Steering Committee should make an ongoing effort to convert the entire workforce. Eventually all employees will have to be advocates. Those who cannot make the transition will have to be removed. However, it is always better to convert than to remove.

=========== SUMMARY ===

1. One kind of team is the *natural work team*. Members of such a team work together every day in the same department or unit. Another kind of team is the *cross-*

Figure 14–1
Team Member Checklist

Mandatory Characteristics

☐ Total Quality Advocate?

☐ Possess Desired Technical Skills?

☐ Unbiased in Outlook?

☐ Non-Volatile Temperament?

Important Characteristics

☐ Suitable Interpersonal Skills?

☐ Possess Desired Experience?

☐ Brings Unique Viewpoint?

☐ Open-Minded?

☐ Possess Good Communication Skills?

☐ Team Player?

functional team. Members of this type of team come from different departments. Total Quality will ultimately require both types of teams.

2. Team composition should be established by the Steering Committee to ensure that teams work on projects that support the organization's vision, and that the teams have the best mix of employees.

3. When establishing cross-functional teams, the Steering Committee should decide, first, what type of individuals are needed and, second, who the individuals should be. There is a natural tendency to want to ask department managers to serve as team members, but this is not the best approach. It is the employees, not the managers, who are most familiar with the process.

4. In deciding who the actual members of a team should be, Steering Committee members should develop a checklist of desired characteristics. Mandatory characteristics are that candidate is a Total Quality advocate, has the necessary technical skills, is unbiased in outlook, and has a nonvolatile temperament. Other attributes that are necessary to the specific project should be identified and included on the checklist.

═══════ KEY TERMS AND CONCEPTS ═══════

Advocate

Cross-functional team

Foolproof mechanism

Natural work team

Poka-Yoke

Team composition

Team member checklist

REVIEW QUESTIONS

1. Explain the rationale for using cross-functional teams.
2. Why is it important for the Steering Committee to select team members?
3. Describe what can happen if early teams contain members who are not advocates.

ENDNOTES

1. Kiyoshi Suzaki, *The New Manufacturing Challenge: Techniques for Continuous Improvement* (New York: The Free Press, A Division of Macmillan, Inc. 1987), 98.

CASE STUDY 14–1

Team Composition at MTC

The MTC Steering Committee now had two initial teams to staff. The first team would be a natural work team in the metal fabrication department. Its task would be to reduce the amount of metal scrapped during fabrication processes. The problem with establishing this team would be deciding who would *not* serve initially. With the exception of one senior machinist who would be assigned other duties, every member of this department was a strong advocate of Total Quality. Applying all applicable criteria, the Steering Committee established an eight-member team to make recommendations for reducing waste.

The second team would be cross-functional. Its task would be to improve internal-customer relations between the Chassis and Circuit Board Departments. The Chassis Department produced the chassis for the various types of electromechanical equipment in MTC's product line. The Circuit Board Department assembled the circuit boards that were placed inside the chassis. The Assembly Department actually put the boards in the chassis. Whenever an assembly problem occurred, the Circuit Board Department blamed the Chassis Department and vice versa. As a result, relations between the two departments were often strained. The Assembly Department was like a Ping-Pong ball bouncing back and forth between the two departments.

The Steering Committee decided to form a team with representatives from the Chassis, Circuit Board, Assembly, and Design Departments. The representative from Design was there to give the designer's viewpoint concerning any changes recommended. Steering Committee members agreed that the Design Department's representative should be an advocate and expert in design for manufacture and design for assembly. To promote intellectual diversity, a representative from Marketing and another from Accounting were added.

CASE STUDY 14–2

Another Surprise at ESC

All Steering Committee members attended the next meeting, out of fear that Lane Watkins might do something that would reflect badly on them. Watkins smiled to himself as he surveyed the faces that stared back at him. "They don't know whether to follow me or stage a mutiny," he thought. In fact, several had tried to do just that by going to the boss, but, just as Watkins had predicted, they couldn't get an appointment. "Well, at least I finally have their attention," Watkins said to himself. And that he did. When he began to speak, the Committee members were all ears.

"Because our first project crosses all departmental boundaries," explained Watkins, "we will need a cross-functional team." When the representative from Civil Engineering asked who would be on the team, Watkins dropped another bomb. "We will," he said.

"Are you saying that this Steering Committee is going to serve as a cross-functional team to improve internal-customer relations?" asked the Accounting representative. "That's exactly what I'm saying," Watkins replied. He went on to explain that they weren't a Steering Committee and never had been. Their respective bosses should have constituted the Steering Committee. However, they would make a perfect cross-functional team. First, they represented every major department in the company. Second, each of them was the closest thing to a Total Quality advocate his or her respective department had.

The Marketing representative summed up the feelings of the group when he said, "Lane, you are either a genius or a nut." Watkins laughed and thought to himself, "At this point, I don't think it matters much one way or the other."

Train the Teams

RATIONALE FOR TEAM TRAINING

The new general manager of a $200,000,000 business once told us that employees didn't need training in Total Quality. He thought that all employees could pick up the concepts and techniques if they had a couple of dedicated leaders. His view of Total Quality was that it amounted to nothing more than common-sense management. It was no surprise to us that this individual had overseen the total failure of one business. What was surprising was that he got a second chance with another, similar-sized business. When it became clear that history was going to repeat itself, he was removed.

This "common sense" definition comes up frequently whenever people talk about Total Quality. In a way, they have a point. Common sense does tell us that it is better to have satisfied customers than unsatisfied ones. It is also common sense when we say that it is better to make a product right in the first place than to spend more money correcting it. But if Total Quality were simply common sense, why did it take the world so long to catch on? Management worldwide worked at cross-purposes with what we now call the Total Quality philosophy for decades until Deming and Juran came on the scene. Where was common sense before Deming and Juran? Perhaps this is what Kaoru

---

TOTAL QUALITY TIP

Training Is Critical

The consequence for not training your Huns is their failure to accomplish that which is expected of them.[1]

Wess Roberts

Ishikawa meant when, reflecting on his early years, he said that Japanese industry and society behaved very irrationally. He went so far as to call Total Quality a "thought revolution in management."[2]

Actually, the common-sense argument misses the point. Total Quality does appeal to common sense, but common sense has had little to do with traditional management methods. We find that people who believe there is nothing more to Total Quality than common sense usually have a very superficial understanding of its principles, if they understand them at all.

Training is one area in which the common-sense argument tends to break down. It is common sense that people will need training before they undertake something entirely new. Yet far too many organizations fail to acknowledge the importance of training.

When a team of employees prepares to begin their first Total Quality project, they typically lack the skills necessary for success. Employees assigned to teams may not be accustomed to making decisions because, in the past, managers made all the decisions. Traditionally, managers made decisions because it was felt that they were the only ones who had the education and experience needed to make sound decisions. In the typical Western organization, employees did the work and management made the decisions. The goal of Total Quality, however, is to push decision making down to the lowest possible level. This is an entirely new concept for many employees, and nothing in their experience has prepared them for it. Obviously, they need training. They must develop the problem-solving tools and the interpersonal communication skills needed in order to be effective team members. If team members don't have these skills, failure is virtually assured. In a Total Quality organization, nothing is more important, more essential, or more critical to success than team training.

SUBJECTS TO BE COVERED IN TRAINING

Every new team should receive training in Total Quality concepts, teamwork, communication, the Plan/Do/Check/Adjust (PDCA) cycle, and the basic Total Quality tools. These topics represent the entry-level minimum for team training. Depending on the team's assignment, more may be needed. For example, when Harris Semiconductor decided to implement *self-directed work teams* on the factory floor, it was necessary to provide training for employees in physics, chemistry, mathematics, and electrical circuit analysis, in addition to the topics listed above.

Working with a local community college, Harris trained its team members by sending them to class four hours a day, five days a week for one year on company time. Harris's program is ongoing as more and more self-directed work teams are formed, and is widely recognized as a best-in-class system.[3] This is not to say that your organization must necessarily invest that heavily in training, but that specific situations may require training beyond entry-level basics.

Our intent here is to list the specific subjects to be included in most team training efforts. In Steps 3 and 4, we discussed team building and Total Quality training for the Steering Committee. It may be helpful to review those chapters. Team training will be similar, although reduced in depth.

Total Quality Concepts, Teamwork, and Communication

Chapter 1 of *Introduction to Total Quality* provides the information needed to instruct the team on basic Total Quality concepts.

A few hours invested in teamwork training will pay dividends. All new teams should receive instruction in the principles and rules of teamwork and communication. For many employees, working as a team will be a new and different experience. It is not something that comes naturally. Chapters 8 and 10 of *Introduction to Total Quality* provide the substance of the information to be conveyed.

We believe it is possible to conduct this element of training in half a day. However, a full day would not be considered too much. The amount of detail you plan to include will determine the duration and cost of the training.

The Plan/Do/Check/Adjust (PDCA) Cycle

According to Dr. W. Edwards Deming, the PDCA cycle was the most valuable concept he taught the Japanese. Dr. Deming told Japanese industrialists repeatedly that by using the PDCA cycle they could improve the quality of their products sufficiently to compete in world markets, and then continually improve even beyond that level. PDCA is just as applicable in Western organizations as the driver of continuous improvement and customer satisfaction.

The PDCA cycle was explained in Step 12 and illustrated in Figure 12–2. Since the logic of the PDCA cycle is not intuitively obvious, we explain it again here:

Step 1: Plan   When planning a product, we want input from potential customers as to what features they think are most desirable. When planning a process, we want to learn what the customers of the process need. When planning a team project, we want to learn what the team is supposed to accomplish. In each case, we have to determine which data are available, which additional data will be needed, which specific steps must be taken, and in which order. This is the planning element of PDCA.

Step 2: Do   Carry out the plan.

Step 3: Check   Observe what happened when the plan was put into action. What were the effects of what was done in the previous step? Collect the data needed to determine whether the action taken produced the desired result.

Step 4: Adjust (or *Act*)   Study the results. If they are less than expected, go back to Step 1, adjust the plan appropriately, and try again.

Repeat the PDCA cycle continually for as long as the product is made, the process is used, or the project exists. This process is the heart of continuous improvement.

The PDCA cycle will be invaluable to all components of the organization as the implementation moves forward. It is a powerful decision-making tool as well as an excellent catalyst for process or product improvement. PDCA will be used by the Steering Committee to set up projects and establish teams; by teams to guide their activities; and by teams and the Steering Committee to ensure that improvement activities are producing the intended results.

Problem-Solving Tools

The remaining training should be devoted to problem-solving tools. These tools were introduced in Step 4 and illustrated in Figure 4–3. You will find more detailed information in Chapter 12 of *Introduction to Total Quality*. The nine basic tools of Total Quality are:

1. Pareto Charts: to separate the important from the trivial.
2. Fishbone Charts: to identify and isolate causes of problems.
3. Stratification: to group data by common elements so as to facilitate interpretation.
4. Check Sheets: to facilitate collection and interpretation of data.
5. Histograms: to depict frequency of occurrence.
6. Scatter Diagrams: to determine the correlation between two variables.
7. Run Charts and Control Charts: to record the output of a process over time and to separate *special* and *common* causes of variation.
8. Flow Charts: to describe inputs, steps, functions, and outflows so that a process can be understood and analyzed.
9. Surveys: to obtain relevant information from sources that might otherwise not be heard from.

Not every team will use all of these tools. Examine the task being assigned to the team, try to determine which of the tools will be helpful, and train team members in using those tools. Be sure to restrict training to only those tools that are needed. For example, it would be pointless to teach flow diagramming to a team that won't use flow charts in their project. If the need to use flow charts arises later on, the team should receive the necessary instruction at that point. Relevance and immediate applicability will make instruction more meaningful.

In special cases, teams may need even more specialized tools and techniques. In such cases, you may require assistance from outside the organization. For example, Design of Experiments (DOX or DOE) is one such specialized, highly sophisticated tool that would not be included in the basic instruction curriculum.

A team will need a four- to eight-hour class on the nine basic tools listed earlier, or a subset of those. The idea is to give teams the basic instruction, and then let them learn by doing. Teams should be led or facilitated by a leader/facilitator who has a good grasp of the tools and will be able to help team members who have trouble applying the tools properly.

TIMING OF THE TRAINING

When the Total Quality movement was getting started in the United States back in the mid-1980s, a popular method for measuring progress was simply to count heads. The success of Total Quality was measured by counting the number of people that received Total Quality training, the number of teams deployed, and so on. This method prevailed because it was difficult to accurately measure the results of Total Quality, and, as always, top management felt that numbers were the only way to show that progress was being made. Unfortunately, this type of measurement meant little in terms of real progress.

Because training is not retained unless applied immediately, training employees whether they have an immediate need for training or not accomplishes little more than increasing training expenditures. Organizations soon learned the error of their ways. Since all employees could not be assigned meaningful team activities all at once, many forgot what they had learned and, as a result, training dollars were wasted.

We now understand the training should be given on a just-in-time basis. The training should be provided when a team is formed for a specific project. There are two major advantages to this approach. First, the training will be fresh when it is applied in a real setting. Second, the training can be tailored to the specific task assigned to the team. This will reduce the cost of the training to the least possible amount that is consistent with the team's task, and prevent wasting time and resources on training a technique that has no application.

TEAM TRAINING AS A STEERING COMMITTEE FUNCTION

It should come as no surprise that training is the responsibility of the Steering Committee. Since the Steering Committee is the senior management entity in a Total Quality organization, it is the Committee's responsibility to provide training. Steering Committee members may or may not actually do the training. The Steering Committee should identify the training approach best suited to the organization's needs and see that the training is provided.

APPROACHES TO TRAINING

Common approaches to training include cascaded training, training by designated in-house trainers, and training by outside consultants.

Cascaded training is often an effective method for deploying training through the organization. This approach involves passing the training down from senior management to middle management and from middle management to the rest of the employees.

Recall that in Step 4, the Steering Committee received Total Quality training. With cascaded training, a member of the Steering Committee who has received training is asked to invest some time in training middle managers; middle managers then train their subordinates, and so on until all who need training have received it. Cascaded training has several advantages:

1. People are likely to work harder at learning if they expect to teach the subject matter to others.
2. Preparation for teaching and the actual teaching itself reinforce the teacher's knowledge of the subject.
3. When managers teach, it shows they are serious about the subject matter.

On the other hand, cascaded training does have some disadvantages. The most important of these is that not everyone is an effective teacher. However, the cascaded training approach has sufficient advantages to warrant adapting it to the just-in-time training philosophy. This can be done by having one or more members of the Steering Committee train the teams as they are formed, or by having training done by one or more middle managers who have already received their training from a Steering Committee member.

Another approach is to train designated trainers and then have them train the teams. This is the most widely used method of deploying Total Quality training. It has the advantage of consistency of instruction. If trainers also act as facilitators for the teams, the training can be continued indefinitely as on-the-job training.

A third approach is to use an outside consultant to do the training. This may be most useful for the early teams, until an in-house capability can be developed.

Regardless of who does the team training, the person must be competent to do the job. In addition to being completely knowledgeable in the fundamentals of Total Quality, teamwork, and communication skills, instructors must also be good teachers. This means they must know how to prepare, present, apply, and evaluate instruction. Too much is at stake to risk haphazard instruction.

TEAM TRAINING AS AN ONGOING PROJECT

Keep in mind that we are not simply talking about the training of one team. After the first team is activated, another will be formed, and another, and another. There will be a steady stream of teams coming on-line, and each will have to be trained. As teams complete assigned projects and disband, others will be created. After awhile, second- and third-generation teams will be formed, composed of team members from earlier teams. However, teams will have first-time members for a long time, perhaps for a year or more, and these employees will require training. As new employees join the organization, they also will require training. As new tools and techniques are developed, employees will have to be trained in their use. The point is, there will probably never be a time when you can say, "Everyone is trained, so we don't need to train any more." In fact, it seems that the more people learn about Total Quality and its tools and techniques, the more they realize they still have more to learn. We think the best approach is to assume that training will be required forever.

SUMMARY

1. Although the Total Quality approach appeals to common sense, Total Quality is much more than just common sense. Consequently, team training is essential to ensure its success.
2. Team training should cover Total Quality concepts, teamwork, communication, the Plan/Do/Check/Adjust (PDCA) cycle, and the basic Total Quality tools.
3. Timing of team training is critical. The most effective approach is to deliver training on a just-in-time basis. This ensures that the training is fresh when applied. Training teams before they have a meaningful assignment that requires the training wastes resources.
4. Like all aspects of the implementation, the Steering Committee is responsible for team training.
6. One approach to training is cascaded training, which involves training top management first, then having them train middle managers, then having middle managers train supervisors, and so on until everyone has been trained. Other approaches are to train a team of in-house trainers or bring in outside trainers.
7. Team training is an ongoing process. After one team is trained, another will need training. Once all original teams have been trained, new teams will be formed and will require training.

KEY TERMS AND CONCEPTS

Cascaded training	Quality tools
Just-in-time training	Tailored training
Plan/Do/Check/Adjust cycle (PDCA)	Team training

REVIEW QUESTIONS

1. Explain the rationale for providing training for teams.
2. What subjects should be covered in team training?
3. Why is it important to provide team training on a just-in-time basis?

4. Who is responsible for team training?
5. Describe three approaches to training.
6. Explain the concept of cascaded training.
7. Is team training ever completed? Explain.

ENDNOTES

1. Wess Roberts, *Leadership Secrets of Attila the Hun* (New York: Warner Books, 1991), 110.
2. Kaoru Ishikawa, *What Is Total Quality Control? The Japanese Way* (Englewood Cliffs, NJ: Prentice-Hall, Inc., 1985), 103.
3. Steve Gilmore, Ed Rose, and Ray Odom, "Building Self-Directed Work Teams," *Quality Digest,* December 1993, 29-33.

CASE STUDY 15–1

Team Training at MTC

John Lee and the rest of MTC's Steering Committee spent more than an hour discussing how to handle team training. Eventually they decided to adopt the cascaded training approach. Each member of the Steering Committee had undergone extensive training during which a comprehensive package of training materials had been collected. All members of the Steering Committee wanted to train their mid-level managers and then cascade the training throughout the organization. Lee made the outside facilitator available to the Steering Committee members to assist in developing their respective training packages.

The two teams that had already been selected would be the first to be trained. In both cases, the same five topics would be covered: Total Quality concepts, teamwork, communication, the Plan/Do/Check/Adjust cycle, and Total Quality tools. The outside facilitator worked closely with the two Steering Committee members in developing their training plan and supportive materials. The Steering Committee members were shown how to use the four-step teaching approach: preparation, presentation, application, and evaluation. The training sessions were monitored by the facilitator and evaluated by participants. With input from the facilitator, the instructors were able to improve the quality of their instruction as they went.

Much was learned that would help other Steering Committee members in preparing, presenting, applying, and evaluating their instruction.

CASE STUDY 15–2

Team Training at ESC

Lane Watkins knew he couldn't bring in an outside facilitator to provide team training. But who could he get to train ESC's first cross-functional team? Nothing that had happened thus far had fit what he knew about Total Quality. By now the Steering Committee should have been thoroughly trained in all aspects of Total Quality. Its members could then serve as in-house instructors for providing team training. Such an approach would have allowed ESC to cascade training.

Unfortunately, cascading was not an option. In fact, Watkins wasn't sure he had *any* options. The Steering Committee had not even been trained. Consequently, its members could not serve as instructors. In fact, all of the Steering Committee members were on the team that needed to be trained! Watkins couldn't bring in an outside facilitator because the implementation itself hadn't been approved by higher management. "What a convoluted mess!" thought Watkins.

There was one possibility, but it certainly wasn't appealing. Watkins could train the team himself. He didn't like the idea for a variety of reasons, but at this point he had painted himself into a corner. Under the circumstances, providing the training himself was his only option.

Watkins decided to stick to the basics. Team members would learn to use Total Quality tools, work as a team, communicate better, and apply the Plan/Do/Check/Adjust cycle. The actual training was conducted as four-hour sessions spread over a week. To his surprise, Watkins found that the training sessions went well. The team members enjoyed the topics that were covered and saw numerous potential applications at ESC. "That's the problem," thought Watkins. "These people want to succeed and they have the talent, but higher management is in the way."

The first Total Quality team at ESC had been trained. Watkins felt relief, but was concerned about what might lie ahead. Now that the team was trained, its members would want to take on a project. Watkins could think of no way to take the next step without risking the wrath of higher management. "Well, what the heck," he thought. "I've come this far, I might as well go all the way." Even so, Watkins resolved to check his mailbox that evening, in hopes there been some response to the resumes he had sent out.

Activate the Teams

The twenty-step implementation process illustrated in Figure I–12 shows three phases for the complete implementation process: preparation, planning, and execution.

The preparation phase included the first eleven steps, from obtaining commitment from top management through establishing baseline levels of customer satisfaction. During that phase the Steering Committee was formed and trained. The organization's vision, supporting broad objectives, and guiding principles were documented and communicated to all employees. The organization then looked at itself objectively and identified its strengths and weaknesses, advocates and resisters, and baseline levels of satisfaction for both its employees and customers. All of this was done in preparation for the next phase: planning.

The planning phase involved Steps 12 through 15. During this phase, the Steering Committee laid out the approach for the organization's entry into Total Quality. Implementation was tailored to the special attributes and needs of the organization, initial projects were identified, and teams were formed and trained. All of this was done in preparation for the third phase: execution.

With Step 16 we now enter the final phase of implementation: execution. In this phase we will activate teams to undertake their assigned projects. In addition, we will use the PDCA cycle to establish a closed loop between project teams and the Steering Committee. After enough time has passed for a change in employee and customer attitudes to have taken place, we will take another look at the projects to make sure Total Quality

TOTAL QUALITY TIP

Successful Teams Must Be Nurtured

"Well-functioning teams are created, not designated . . . composite teams require significant amounts of energy to guide their formation—much more than sending the new members to a session or two of team-building activities, and then calling them a team!"[1]

Robert M. Tomasko

is having the desired impact. Finally, we will make changes to the organization's infrastructure to eliminate roadblocks and to start converting the organization to the ideal Total Quality state.

From this point on, Total Quality shifts from implementation to a planning and execution cycle. Steps 12 through 20 become a continuous process. Some of the steps from the preparation phase, particularly Step 7, "Communicate and Publicize," will also be continued forever, in one form or another. The Steering Committee will periodically return to Steps 5 and 6 to ensure that the organization's vision, guiding principles, and broad objectives remain appropriate in a changing world. Clearly, the Steering Committee will maintain an emphasis on Step 15 with continued training. Now, in Step 16, Total Quality goes to work for the organization and all of its stakeholders.

RATIONALE FOR FORMAL TEAM ACTIVATION

Now that the Steering Committee has identified a project, established a corresponding team, and trained the team members, it may seem that all that remains is for the Steering Committee to tell members what is expected and turn them lose. Not so! There is much more to be done. A major cause of failure of Total Quality is team members' incomplete understanding of what they are supposed to do, the boundaries within which they are to work, how much time they are to spend, when or if they are to report progress or problems and to whom, when they are expected to be finished, and what authority they have. These are not questions that teams can answer for themselves. The Steering Committee must assume this responsibility by establishing team charters at the outset. Otherwise, drifting, ineffective teams will result, and the organization's Total Quality effort will fail to produce the desired results.

DEVELOPING THE TEAM CHARTER

Developing the team charter is the responsibility of the Steering Committee and should flow naturally from the project-identification process. The purpose of the team charter is to document exactly what the Steering Committee expects of the team. When a team is assigned a project, it is not always clear to team members how far they are expected to go. The charter answers this question.

For an example, we return once again to the company that manufactures designed-to-order power supplies used in military and space applications. One of its initial projects was to look at the process involved in moving a product from contract to production, with the objective of reducing the time and cost of the design and prototype phase. The company's management team had expected the project team to determine what happens from the point where a contract is awarded, through the engineering phase, to the point where the design package is given to the Manufacturing Department for production. The problem with the assigned task was that it was so broad that the team found it difficult to focus. Team members believed strongly that the biggest problem was that the Engineering Department wasn't involved in the early phases of contract negotiation, and as a result did not know what the customer wanted. Since early involvement by Engineering was the easiest place to make quick and significant savings, the team decided to pursue that goal. However, their recommendation got a less-than-enthusiastic response from the management team because it felt that the project team, by pursuing the issue of contract negotiations, had transgressed the boundaries of its charter. As a result, the team lost several weeks of work and became demoralized, nearly to the point of giving up. Had a formal charter with clear boundaries been established, this would not have happened.

This example illustrates how vital it is that the Steering Committee provide a charter to each team. The team charter should be a written document, to ensure complete and uniform understanding among all team members. The charter will be used by the team to develop its own mission statement and to settle any territorial questions that come up during the life of the team.

CONDUCTING THE TEAM ACTIVATION MEETING

Is the written charter all that is needed to put a new team on the intended path? No. While the written charter is vital, it is not sufficient to launch a team. Typically, teams are composed of five to seven people who come from different backgrounds. Their levels may run the gamut from the bottom of the hourly labor ranks to high-level professionals. Hence, there must be frequent interaction between the team and the Steering Committee. The Steering Committee's expectations in terms of project outcome, reporting, use of special techniques such as the PDCA cycle and certain of the Total Quality tools, time available, and responsibilities and authority must be clearly articulated and understood. This should be accomplished in a formal team activation meeting involving the Steering Committee and the project team. All questions from the team should be answered during this meeting. At the end of the meeting, all Steering Committee and project team members should understand what the team is to do, how much time is to be spent, the team's operating boundaries, and all other pertinent information.

Published minutes of the meeting should reflect all critical instructions and issues so that both the project team and the Steering Committee can refer to them as necessary over the life of the project. The written charter and the minutes of the team activation meeting will be used by the project team in drafting its mission statement.

Instructions that should be given to the team during the team activation meeting are illustrated in Figure 16–1 and discussed in the following paragraphs.

Overview of the Project

The team should be informed of the thinking that led the Steering Committee to assign the project. What is the problem or issue in question? How does the project fit in with the organization's vision and broad objectives? What are the Steering Committee's expectations for the project? The team should also be informed of the Steering Committee's thinking related to team composition—why team members as individuals were selected for the team and what each is expected to bring to the team. Depending on the

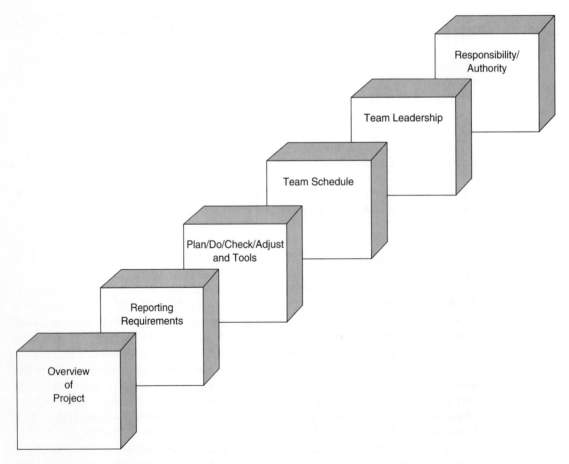

Figure 16–1
Items to Cover when Giving Instructions to a Team

kind of team (i.e., natural work group or cross-functional) the information conveyed may vary, but the object is to make the team as fully aware of the issues, thinking, and expectations of the Steering Committee as possible. For example, if competitive pressures or financial considerations underlie the project, project team members should be told. You cannot give them too much information.

Reporting Requirements

Teams should be told how often they should report and to whom. A formal reporting procedure should be set up for every team, and the content and timing of reports should be specified. It is a good idea to maintain a tight feedback loop at first, perhaps loosening it up a bit once team members are working well together and are on the right track. At first, the team should provide reviews of its activities by giving live presentations to the Steering Committee every two weeks. After two or three of these, if the team is doing well, the interval can be lengthened to once a month. The review presentations should be made by team members. The presentation need not be so formal or structured that preparation requires a lot of time. All that is needed initially is for the Steering Committee to get a feel for how the team is working and whether progress is being made. When the team is ready to make its recommendation(s), it should thoroughly prepare its presentation, including handouts showing the salient points and how the recommended solution will improve the process or solve a problem.

The Plan/Do/Check/Adjust Cycle and Total Quality Tools

All team activity should make use of the PDCA cycle, but this may not be so obvious to the early teams. Consequently, it should be emphasized. The Steering Committee should let the team know how they want the project run, emphasizing the PDCA cycle and perhaps suggesting some particular problem-solving tools that might be useful. At the same time, it should be understood that, except in the most general terms, the team is to manage its project.

Team Schedule

The Steering Committee has a difficult task here. A schedule for the team's project must be provided, but it should be communicated to the team as a goal rather than as a rigid timetable. The team should set its own schedule, using the Steering Committee's input as guidance. Major discrepancies should be discussed with the Steering Committee at the beginning of the task, or whenever a discrepancy becomes apparent. As the team gets into its work, unexpected discoveries or events will undoubtedly occur, and some may require that the team be given more time. It is also possible that the team may conclude its project in a shorter time than the Steering Committee anticipated. Such deviations from the Steering Committee's schedule estimate must be accommodated. The Steering Committee will get input during regularly scheduled reports from the team and may judge then whether the team is making satisfactory progress or whether it needs a gentle nudge or a course correction.

TOTAL QUALITY TIP

See Your Organization through the Eyes of the Customer

"Every leadership team needs to look carefully at the organization through the eyes of its customers, and see what they see. They also need to look at the ability of . . . people to work together internally."[2]

Karl Albrecht

Team Leadership

Teams generally should designate their own leader. Having the Steering Committee select the team leader conveys the message that the team is being run by management. In some situations the leader will be designated by management, but for the most part, the selection of team leader is best left to the team. We recommend that the Steering Committee instruct the team to select a leader. The team leader's responsibilities will be to conduct meetings, assign action items for team members, and communicate with the Steering Committee as required between formal reporting dates. The leader is also responsible for having someone record the minutes of each meeting; the minutes should cover all action items assigned and the due dates for those items. The leader should publish an agenda for each team meeting and distribute it to the team a day or two in advance. There will be other duties that naturally fall to the leader. Of course, team leaders may delegate to the team members as appropriate.

Responsibilities and Authority of Teams and Team Members

Steering Committees must clearly define the team's responsibilities and authority. For example, in an earlier step we discussed a team that was chartered to solve a problem with customer service. The team's responsibility could have been to find the problems in the processes and change the processes to make them foolproof. Or, its responsibility might have been to analyze the processes, find the problems, and make recommendations to the Steering Committee, which would then use its authority to change the processes.

Normally, even cross-functional teams do not carry the necessary authority to make changes that cross departmental boundaries. Even if boundaries are not a question, limiting the team's responsibility to providing *recommendations* ensures appropriate checks and balances. Bringing the recommendation for a process change to the Steering Committee ensures that it will be discussed by another set of people. If the recommendation is adopted, the approach also ensures management buy-in. Regardless of how much authority is given to a team, it should be clearly defined by the Steering Committee at the outset.

The responsibilities and authority of team members should also be clearly delineated. Team members are responsible for attending scheduled meetings. In addition, they must have the authority to spend additional time working on action items. The Steering Committee must ensure that the supervisors of individual team members give them the time needed. For example, if an action item requires a team member to interview other employees, the time and authority to do so must be provided by the Steering Committee. Generally speaking, a team member has the authority to act within the boundaries of the charter and the team's instructions. From time to time, issues will surface that call for the Steering Committee to step in to provide the necessary authority, or at least to make it clear that the authority does exist.

PROVIDING A TEAM FACILITATOR

The Steering Committee should provide a trained facilitator for the team. The role of facilitator is not to act as team leader, but to keep the meeting focused and progressing. The facilitator will be able to provide help with the selection of problem-solving tools and ensure that they are being used appropriately. Another important function of the facilitator is keeping the discussion moving while preventing it from being dominated by an individual team member. The facilitator should also sum up discussion and bring it to a close when it appears to have run its course. The agenda should allocate a reasonable amount of time for each item, and the facilitator should see to it that discussions do not exceed the allotted time.

For the initial teams, it is important that a well-trained facilitator attend every team meeting and be available for consultation between meetings. Many organizations use their trainers as facilitators, an approach that usually works well. As experience is gained, use of a facilitator may be discontinued, with that role given to team members on a rotating basis.

COMMUNICATION AND PUBLICITY

As teams are activated, there should be sufficient communication and publicity to ensure that the entire organization knows the task assigned to the team and what that task is expected to do for the company and its employees and customers. All employees should know who the team members are, in case the team needs to call upon their expertise. Communication activities are a signal to supervisors and managers that team members are on a mission from top management and that help should be provided when needed.

=========== SUMMARY ===========

1. Team activation is a critical procedure. Team members need to know what is expected of them, when they should complete their responsibilities, and what authority they have to act. Communicating this information is the purpose of formal team activation.

2. The team charter is a document that clearly sets forth the team's purpose and what is expected by the Steering Committee. It is presented in written form to rule out any chance of a misunderstanding.

3. The team activation meeting allows the Steering Committee to instruct the team in the following areas: overview of the project, reporting requirements, the Plan/Do/Check/Adjust cycle and Total Quality tools, schedule expectations, team leadership, and team responsibilities and authority.

4. All initial teams should have a trained facilitator. The facilitator's job is to keep discussion flowing, keep meetings on track and on time, and serve as a technical advisor to the team.

5. Once a team is formed, its members' names and its responsibilities should be publicized. All employees need to know that the team is on a mission from top management and that complete cooperation is expected.

KEY TERMS AND CONCEPTS

Authority	Team activation
Communication and publicity	Team activation meeting
Overview of the project	Team charter
Plan/Do/Check/Adjust cycle	Team facilitator
Reporting	Team schedule
Responsibilities	

REVIEW QUESTIONS

1. Explain the rationale for formal team activation.
2. What information should the team charter contain?
3. Why is the team activation meeting important? What should be accomplished during the meeting?
4. Explain the instructions that should be given to the team in each of the following areas: overview of the project, reporting requirements, the Plan/Do/Check/Adjust cycle and Total Quality tools, schedule, and team leadership.
5. Differentiate between the authority of the team and the Steering Committee.
6. Describe the responsibility and authority of individual team members.
7. What is the role of the team facilitator?
8. Why are communication and publicity important?

ENDNOTES

1. Robert M. Tomasko, *Rethinking the Corporation: The Architecture of Change* (New York: Amacom, American Management Association, 1993), 92.
2. Karl Albrecht, *The Northbound Train* (New York: Amacom, American Management Association, 1994), 93.

CASE STUDY 16–1

Team Activation at MTC

John Lee and MTC's Steering Committee had two team activation meetings to conduct. One was to activate the team that had been selected to make recommendations for reducing waste in the Metal Fabrication Department. The other was to activate the team that would attempt to improve internal-customer relations between the Chassis and Circuit Board Departments. The Steering Committee decided to activate the Internal-Customers Team first.

After developing a written charter, the Steering Committee called the team in for the activation meeting. Team members were given copies of the charter and encouraged to ask questions for clarification. Their instructions included a discussion of the project reporting requirements, use of the Plan/Do/Check/Adjust cycle, a tentative schedule, selection of a team leader, and an overview of the responsibilities and authority of the team and of individual team members.

Several team members expressed concern about whether they would be given the time they would need to carry out individual tasks. John Lee gave team members a copy of a memorandum he had already delivered in person to their respective supervisors. That satisfied their concerns. The team was instructed not to take unilateral action, but to present recommendations to the Steering Committee, which, in turn, would take responsibility for having them implemented.

The team members seemed to be pleased with the Steering Committee's choice of a facilitator. Once the facilitator had been introduced, the team members departed to select a leader and begin their preparation activities.

CASE STUDY 16–2

Team Activation at ESC

Now that a project had been selected—or rather, now that Lane Watkins had chosen one—the Steering Committee members were ready to take on their roles as members of a project team. Watkins had developed a written charter and was prepared to give the team its instructions. However, he had barely gotten started when his choice of the Steering Committee as the first project team came back to haunt him. Several team members didn't agree with the charter Watkins had drafted, and they began suggesting changes. Since they were also Steering Committee members, Watkins had no choice but to go along.

The same problem arose again when he gave the team its instructions. Any time team members disagreed, they simply stepped out of their role as project team members and into the role of Steering Committee member. Worse yet, Watkins found himself cast into the role of the facilitator rather than the team leader when the team selected the leader by secret ballot. Now Watkins had to contend not just with a lack of commitment on the part of higher management, but with a renegade project team. This was not a welcome turn of events.

When the new team leader initiated a discussion of the team's mission and schedule, it became clear to Watkins that the real mission was to stall and hope that the idea of Total Quality at ESC would simply die a quiet death. Watkins saw that he had been outfoxed by his colleagues. "Alright, you win this round," he thought. "But the fight isn't over yet!"

Provide Team Feedback to the Steering Committee

- Rationale for Feedback to the Steering Committee
- The Feedback Loop as Part of the PDCA Cycle
- Frequency and Format of Reporting
- Feedback from the Steering Committee to Project Teams
- The Feedback Loop and PDCA Cycle as an Ongoing Process

This step is the second in the execution phase of the implementation. It has been our experience that it is easy for management to relax after activating teams. Consider what has been done up to this point. The Steering Committee has selected a project, named a team, trained the team, and given team members all available information that was relevant to the project. The Steering Committee has explained how the project relates to the organization's vision and broad objectives, and its expectations in terms of a schedule and desired results. At this point, it is easy to assume that the project is in capable hands and to let the team proceed with no further intervention by the Steering Committee. However, there is a great danger in such an approach. The Steering Committee must stay in close contact with the work of its teams, especially in the early stages. Neglect will lead to failure.

RATIONALE FOR FEEDBACK TO THE STEERING COMMITTEE

There are several reasons for requiring frequent feedback from teams to the Steering Committee. The most important of these are as follows:

1. *The Steering Committee is responsible for the organization's performance.* In a Total Quality Organization, the Steering Committee is management, and management is

---

TOTAL QUALITY TIP

Teamwork Is More than Working Together

"There was a time when a work team was little more than a number of people each doing a similar task or even just pulling on the same rope and thus combining efforts to achieve what could not be done by one person. But it takes more than a bunch of people pulling on the same rope to make a team."[1]

Marshall Sashkin and Molly G. Sashkin

---

still responsible for ensuring that all activities promote the achievement of the organization's vision and broad objectives. Said another way, management is responsible for the success of the organization—today and in the future. It is therefore necessary for the Steering Committee to know what the teams are doing. If the teams appear to be deviating from the planned course, the Steering Committee must make the necessary adjustments. This can happen only if the Steering Committee receives periodic feedback.

2. *New teams are unknown entities.* New employees require special attention until they learn their jobs. This is also true of new teams. Even if all team members are experienced, well-known employees, how they will interact as a team will not be known until they have been observed for a while. Periodic feedback allows the Steering Committee to make the necessary observations.

3. *With initial projects, teams lack experience.* By the time the first teams are established, the Steering Committee is well experienced at working as a team, and it is easy to mistakenly project this experience on to a newly formed team. It is a mistake to assume that, even with training, teams will adopt a teamwork mentality automatically. Feedback from teams helps the Steering Committee determine whether additional training or facilitation are necessary.

4. *The Steering Committee must ensure that instructions aren't misunderstood.* Even under the best of circumstances, instructions can be misunderstood. Issuing instructions to an individual requires great care to ensure that the instructions are completely and accurately understood. Issuing instructions to a team is even more difficult, because the instructions can be understood in as many ways as there are team members. Feedback from teams enables the Steering Committee to assess the adequacy of communication.

THE FEEDBACK LOOP AS PART OF THE PDCA CYCLE

In the Steering Committee's Plan/Do/Check/Adjust cycle, team feedback represents the *Check* step. Without this feedback, the cycle cannot be completed. Feedback enables the Steering Committee to determine whether the team is on track. When necessary, the

Steering Committee provides clarifying instructions, a course correction, or some other kind of adjustment, thus closing the feedback loop.

As the project team operates its own PDCA cycle, the Steering Committee operates one too, at a higher level. Don't confuse the two. The Steering Committee's PDCA cycle is as follows:

Plan   Select a project, select and train the team, and develop instructions for the team.

Do   Activate the team and issue instructions in the form of a team charter.

Check   Monitor the team's progress. Team feedback is an integral part of this function.

Adjust   When monitoring indicates that a team is getting off course, or that adjustments are in order, issue new instructions to the team.

From this discussion it should be clear that the Steering Committee's PDCA cycle involves planning by the Steering Committee, putting a team to work on the plan, monitoring the activities of the team, and managing the team. Without the feedback loop, the Steering Committee's PDCA cycle cannot be completed.

FREQUENCY AND FORMAT OF REPORTING

Once the system is established and some experience has been gained, a monthly feedback reporting cycle for all teams is usually sufficient. However, during the early stages, until both the Steering Committee and the teams have developed some successful experience, a two-week or even a one-week reporting cycle is advisable. In this way, the feedback loop is made tighter and the potential for wasted effort by a team that goes off in the wrong direction is minimized. Teams are more receptive to changing direction before they have dedicated a lot of time to a given approach.

To illustrate, let's return to the example we used in Step 16. The project team was assigned the task of looking at the process involved in moving products from contract to production, with the objective of reducing the time and cost of the design and prototype phase. Initially, the Steering Committee did not include the CEO and several department heads—a critical weakness. Further, the Steering Committee had little experience, as reflected in the charter and instructions given the team. Early discussions in team meetings convinced team members that most of the wasted time and cost resulted from the way in which contracts were pursued, negotiated, and accepted. The team members believed that if certain procedures were changed for the phase between pursuit and award of a contract, significant time and cost savings could be realized for the phase between award of the contract and production of the product. The team members spent several meetings developing these procedural changes before making their recommendations to the Steering Committee.

Although Steering Committee members saw the wisdom in the team's recommendations, they were concerned because the recommendation affected the Marketing and Contracts Departments, which were not represented on the Steering Committee and were not involved at the management level in the work of the team. Moreover, the team

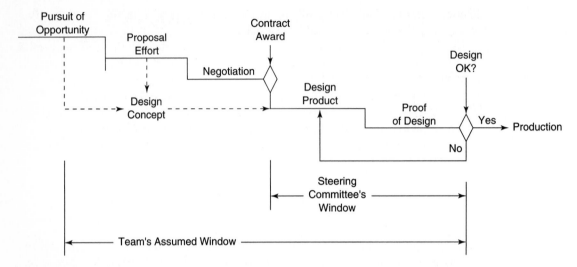

Figure 17–1
Differing Perceptions of Project Team and Steering Committee about Scope of Project

had gone beyond its charter. The Steering Committee had intended the team to look at opportunities for improvement in the window between the award of a contract and the point where production started. The team saw a major opportunity in advance of contract award and went after it. Figure 17–1 illustrates the differences in perception between the project team and the Steering Committee.

Having spent several weeks on the project, the team members became dejected when the Steering Committee told them they were out of line. Several members of the team quit, and the rest lost hope of making a real difference for the company. Eventually the company did implement the changes the team recommended, and the team, by then infused with new members, went back to work. But the example illustrates the problems that can occur when a new Steering Committee does not have authority over all of the departments; when a new team has virtually no experience; and when instructions to the team are poorly developed. The biggest mistake a Steering Committee can make is failing to require the team to report back with its goals and plans before proceeding. Had this been done in the example just described, the team's direction could have been altered before so much time had been spent working outside the realm of their activation instructions.

Figure 17–2 contains recommendations for frequency of progress reports from teams. It is important that the frequency of feedback be at least once per month. In addition to the reasons already given, interacting with the Steering Committee lets team members know that what they are doing is important to management. The format of the feedback reports is also important. Although some team members may not be accustomed to making stand-up presentations to top managers, this is something that can be quickly learned. We recommend that such presentations be required. The benefits far

---

TOTAL QUALITY TIP

Top Managers Must Stay Involved

"If you're going to have an initiative, you've got to create a team of senior managers to coordinate it and to provide the energy. You're moving from a standing start, and that always takes more energy than keeping something going. Unless it starts at the top, it has great difficulty getting started because other priorities take over."[2]

Joseph M. Juran

---

outweigh any perceived disadvantages. First, oral presentations offer the opportunities for dialogue to ensure common understanding and to resolve any issues that may arise. Second, the team will know that their work is something that management takes seriously. Third, the experience gained by team members in preparing and presenting reports will contribute to their personal growth.

In addition to verbal material, the team should provide handouts of charts and all written material to Steering Committee members prior to presentations. Members of the Steering Committee may refer to these materials during presentations and use them for a variety of purposes afterward.

Requests and questions should be submitted to the Steering Committee in writing as they develop. For example, questions about the charter may come up from time to time, and such questions must be resolved by the Steering Committee. At times, the team may need the services of outside experts. The team should bring such situations to the Steering Committee's attention at once, and it should do so in writing. This procedure is not intended to preclude dialogue, but rather to make sure that questions are thoroughly understood and receive the attention they deserve.

Feedback Subject	Format	New Teams	Experienced Teams
Team Mission Statement & Goals	Oral and Written	1–2 Weeks	1–2 Weeks
Progress Reports	Oral Presentation	1–2 Weeks	Monthly
Requests/Questions	Written	As Required	As Required

Figure 17–2
Recommended Format and Frequency of Progress Reports

FEEDBACK FROM THE STEERING COMMITTEE TO PROJECT TEAMS

Feedback works both ways. When requests or questions are directed to the Steering Committee, the Committee should respond in the shortest time possible, certainly in no more than one week. If, for some reason this cannot be done, the team should be advised as to when a response will be given and the reason for the delay. We have had hundreds of team members tell us that the primary reason a team gets turned off is that management fails to respond to its recommendations, questions, and requests. This reaction is understandable. People know that managers spend their time on matters they consider to be important. If management cannot find the time to respond to the teams, then obviously the team's activities must not be important to management. When team members are made to feel that their work is not important to management, then they do not want to waste further time on team projects. To prevent this perception, the Steering Committee must respond quickly and sincerely to every request, question, and recommendation.

Team members tell us that no response is the worst response. They usually can accept that a recommendation has been rejected, provided they understand why. Consequently, if the Steering Committee finds it necessary to reject a recommendation from a team, it should advise the team members quickly and explain why the recommendation cannot be adopted. Provided the reasons are understood, there should be no lasting negative reaction from the team. On the contrary, its members probably will be interested in taking a new approach to solving the problem or improving the process.

THE FEEDBACK LOOP AND PDCA CYCLE AS AN ONGOING PROCESS

The feedback loop and the PDCA cycle continue as long as the Steering Committee has teams working. For a Total Quality organization, this means forever. Some Total Quality proponents think that once the process matures, teams will be able to do their work without interacting with the Steering Committee. Our belief is that the Steering Committee, as the Total Quality organization's top management team, has the responsibility to manage. This means nurturing teams, keeping them on track, responding to their needs, facilitating when necessary, and rewarding them for excellence. These functions require that the Steering Committee stay in close touch with the teams, and that interacting with teams be part of the Steering Committee's PDCA cycle. Even in organizations that employ self-directed work teams, the Steering Committee is always aware of the teams' activities. To do less is to abdicate responsibility, and abdication is *not* what Total Quality is about.

=== SUMMARY ===

1. Periodic, ongoing feedback to the Steering Committee is necessary because: the Steering Committee is responsible for the organization's performance, new teams are unknown and unproven, early teams lack experience, and the Steering Committee must ensure that instructions given to teams are understood.

2. Team feedback is the *Check* phase in the Steering Committee's Plan/Do/Check/ Adjust cycle. The Steering Committee cannot complete its PDCA cycle without feedback from teams. The Steering Committee and its teams run their own separate PDCA cycles simultaneously, but the two are not the same.

3. In the early stages of team development, feedback to the Steering Committee should be as frequent as weekly. Later, once a team has proved itself, monthly meetings should be sufficient. In addition to the feedback requirement, interaction with the Steering Committee lets team members know that their work is important.

4. Feedback works both ways. It is important that the Steering Committee respond in a timely manner to recommendations, requests, and questions from teams. It is better to reject a team's recommendation and explain why than to give no response.

5. In a Total Quality organization, feedback and the PDCA cycle are ongoing and permanent. They never stop, because the need to improve never stops.

KEY TERMS AND CONCEPTS

Feedback loop	Questions
Misunderstood instructions	Recommendations
Progress reports	Requests
Project team's PDCA cycle	Steering Committee's PDCA cycle

REVIEW QUESTIONS

1. Explain the rationale for feedback from project teams to the Steering Committee.
2. Explain the relationship of the feedback loop to the PDCA cycle.
3. Differentiate between the reporting requirements of new teams and experienced teams.
4. Explain what is meant by the statement "Feedback works both ways."
5. Why should the feedback loop and the PDCA cycle continue forever?

ENDNOTES

1. Marshall Sashkin and Molly G. Sashkin, *The New Teamwork: Developing and Using Cross-Functional Teams* (New York: American Management Association, 1994), 9.
2. Joseph M. Juran, from an interview in *Training*, Vol. 31, No. 5, May 1994, 36.

CASE STUDY 17–1

Establishing a Feedback Loop at MTC

John Lee and the other Steering Committee members asked the Internal-Customer Relations Improvement Team to report back as soon as its mission and tentative schedule had been drafted. The meeting that resulted had been a good one. The meeting began with a presentation on the team's mission by the team leader. After completing the brief presentation, rather than distributing copies of the team's tentative schedule, the team leader stopped and asked for ratification of the proposed mission. He explained that changes to the mission might result in corresponding changes to the schedule. Consequently, he wanted to get approval of the mission before discussing the schedule.

The team's mission statement was a good one. It reflected a clear understanding of the team's assignment and was written within the limits established by the Steering Committee. It read as follows:

> The mission of this team is to identify ways of improving internal-customer relations at MTC, with special emphasis on the Chassis and Circuit Board Departments.

The mission statement pleased John Lee. It was brief and to the point, yet comprehensive. It showed that team members understood that the main issue was between two specific departments, and that these departments would be the focal point of all team efforts. However, it did not rule out other departments. This was important because the bad blood between the two departments in question might conceivably be caused by processes originating in other departments.

With the mission statement approved, the team leader presented a chronological set of tasks the team would undertake and a projected timetable for completing each task. These tasks included developing an internal-customer satisfaction survey instrument, distributing the instrument to all employees of the Chassis and Circuit Board Departments, collecting and summarizing the completed results, and making recommendations to the Steering Committee.

Several Steering Committee members suggested adding the task of conducting face-to-face follow-up interviews with selected personnel as needed for clarification and insight. The list of tasks and corresponding deadlines was adjusted accordingly and the team's schedule was accepted as revised. The Steering Committee and team members agreed to meet weekly or at the completion of each projected task, whichever came first. When the meeting had adjourned, both groups went back to work feeling positive and confident.

CASE STUDY 17–2

No Feedback, No Loop at ESC

Lane Watkins now saw just how foolish he had been to make ESC's so-called Steering Committee a project team. Now the fox was not just guarding the henhouse, it had moved in with the hens and was sitting on the eggs! Because the project team was also the Steering Committee, all checks and balances had been lost. During the meeting to draft the team's mission statement, the idea that got the most support was:

> The mission of this team is to stall until upper-management forgets this Total Quality nonsense and moves on to the next management fad.

Watkins, playing the role of the facilitator, tried to convince team members to draft a real mission statement and a list of tasks, but he got nowhere. Finally, in desperation, he asked, "If I develop a survey instrument for identifying internal-customer problems, will you at least look at it?" The meeting broke up with a lukewarm commitment from team members to look at the survey instrument Watkins proposed to develop. However, after the meeting, the team leader stopped Watkins in the hall and said, "Lane, we agreed to look at anything you come up with, but that's all. We did not agree to conduct a survey. Remember that."

Collect and Use Customer Feedback

In Step 11 we discussed developing a baseline for customer satisfaction. This was done before Total Quality work had really gotten started. In this step we discuss the subject from the perspective of an organization that is actively engaged in Total Quality work. The reader may wish to review Step 11 before beginning this step.

In the past, many companies assumed that no matter what product they manufactured, and regardless of its quality, the product would sell if enough dollars were pumped into an ever more innovative ad campaign. Corporate arrogance was reflected in a philosophy of "We know what is best for the consumer." Since so many major corporations operated on this premise, consumers had few alternatives. The U.S. automobile industry serves as an example of this approach to doing business. The industry fared well until the mid-1970s, when the Arab oil embargo convinced U.S. consumers to try smaller, more fuel-efficient Japanese cars. Japanese automakers saw an opportunity and seized it by broadening their offerings in the United States. U.S. consumers began to realize how good cars could actually be. At this point, not only did U.S. consumers have options, they had good options. Consumers were not as naive as U.S. automakers apparently thought they were. Given choices, consumers opted for better value.

This scenario was not limited to the automobile industry. The automobile industry is simply the most visible example. Another example is the consumer electronics industry, which has been all but taken over by Japan. In this case, the Japanese got so far ahead in quality, in reduced time to market, and in development of desirable features

---

TOTAL QUALITY TIP

Customer Input Is Critical

"When a company thinks it's so smart that it doesn't have to ask the customer what he or she likes, it ends up with an Edsel."[1]

John Vanderzee, Advertising Manager, Ford Division

---

that products from other countries were simply not competitive. The U.S. consumer electronics industry had dozed off, only to awaken and find that its market had vanished.

Variations on this theme were played out in the United States in industry after industry during the 1970s and 1980s. By the mid-1980s, many companies were beginning to take up the challenge, and quality-improvement programs of all kinds began to spring up. Enlightened managers came to understand that Total Quality was the key to competitiveness because it brought continuous and sustained improvement in products, services, and all other aspects of doing business. Still, it can be difficult to let go of long-held notions and beliefs.

Managers like to think they can sense intuitively whether a strategy is working as intended or not. Some managers think that if they like a product, customers will automatically like it too. Unfortunately, this is not necessarily the case. Many managers have come to realize that what they think of a product is not necessarily relevant. In the final analysis, only one person can decide whether or not a product is acceptable: the customer.

RATIONALE FOR OBTAINING AND USING CUSTOMER FEEDBACK

Since only the customer can decide whether a product or service is acceptable, it makes sense to listen to the customer. Actually, we should do more than just listen. We need to actively seek any and all information the customer is willing to provide about our product or service, and not just after the sale, but during product development. We need information not just about how well a product is manufactured, but also what features customers want added or deleted and how the product can be made more suitable for their needs. Customers have choices, and, as we have seen, will vote with their wallet for the choice that best meets their needs. Whether manufacturing consumer products, offering services, or contracting with the government, we must remember that customers are voting, and to stay in business, we must find out why customers vote as they do and then use that information to get continually better at satisfying their needs.

In Step 11 we established a baseline of customer satisfaction. The reason for doing this before the Total Quality implementation was fully implemented was to use that baseline as a starting point for making improvements to processes, products, and services. Once Total Quality has been implemented, customer-satisfaction information must be obtained on a periodic basis so that it can be used to measure progress. Customers will either validate changes made as the result of implementing Total Quality or they will not. If they do, the organization knows it is on the right track. If they don't, the organization knows that improvements have not been effective. Remember, with Total Quality you manage with facts, not guesswork. Customer feedback gives you facts.

SOURCES OF CUSTOMER FEEDBACK

Most organizations have many sources of customer feedback. These sources have been available all along, but have not always been used to their fullest potential. Sources of customer feedback are illustrated in Figure 18–1 and discussed in the following paragraphs.

Warranty Data

Data from your warranty department can tell a lot about your product. If warranty-related costs are high, product quality is probably low, resulting in frequent warranty claims. Examining the types of claims can yield valuable information. If a particular type of failure is common, the Engineering Department should improve the product's design to eliminate the underlying cause. A team should be established to analyze warranty data on a continuous basis and report its findings to the Steering Committee at regular intervals.

Sales Results

Few organizations look at their sales figures as customer feedback, but this may be the best use of such data. For example, if a typewriter manufacturer held a 20 percent share of the market for years, but recently saw its market share decline to 10 percent, this is a good indication that customers are voting for the competition. Has quality slipped? Have competitors improved their quality? Has the competition launched new models with attractive features? The organization needs to answer questions such as these and act immediately on what is learned. Actually, this should have been done before half of the company's market share was lost.

Organizations should also look at sales figures in dollars, as opposed to just market share. It is possible to maintain market share but experience rapidly declining revenues. In the case of the typewriter manufacturer, assume that a 20 percent market share produced revenues of $20 million two years ago, but the same 20 percent share today produces only $10 million. This indicates an overall market decline. The typewriter market, which had known only growth for more than 100 years, began a steady decline with the advent of personal computers that run word processing software.

The Steering Committee should recognize that sales performance is an indicator of customer satisfaction and should follow sales data closely.

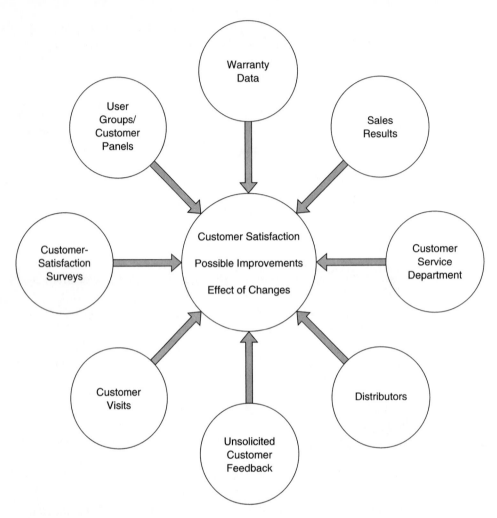

Figure 18–1
Sources of Customer Feedback

Customer Service Department

Companies that employ Customer Service Departments have an excellent source of customer-satisfaction information. Customer service representatives receive a steady stream of customer comments. Often companies ignore this source of information, seeing it as nothing more than complaints. The intelligent approach is to treat customers' complaints as if they were more valuable than gold. Organizations should prefer to listen to complaints from their customers than to have customer voice their complaints to other customers, or even to the competition. Customer complaints provide valuable insights about the organization's products or services. The Steering Committee should receive periodic, objective customer service reports and should analyze them with an eye toward identifying trends, patterns, and customer expectations.

Distributors

Companies that use distributors should insist that all customer feedback, whether positive or negative, be forwarded to the Steering Committee or to a team set up for the purpose of receiving such information. Unfortunately, distributors tend to be effective filters of customer input, blocking most of it. On the other hand, distributors frequently find themselves bearing the brunt of the customer's dissatisfaction. Consequently, they are usually happy to provide feedback if asked.

Unsolicited Customer Feedback

When customers write or call a company to express their disappointment or satisfaction with a product, they usually find themselves talking to the Marketing Department or a public relations representative. In a Total Quality company, customer comments should be directed to the Steering Committee. Be sure that this is the policy in your organization.

Customer Visits

Some companies promote customer visits. Such visits can be an excellent way to collect information on customer satisfaction. An added advantage of the customer visit is that, through dialogue with customers, the organization can clarify the customer's problems or expectations. All such information should be made available to the Steering Committee.

Customer-Satisfaction Surveys

The customer-satisfaction survey is the most common method for soliciting customer feedback. This approach is not new. For years, hotels and restaurants have asked customers to fill out survey cards that provide feedback about service and quality. More recently, it has become common for automobile repair shops to follow up service visits with a telephone call or a mail-in postcard to determine the level of customer satisfaction. Customer surveys can vary from postcards to elaborate multi-page questionnaires that are administered by third-party firms.

The customer-satisfaction survey is a tool that should be used by every organization. Some companies survey a statistically valid random sample of their customers annually. Others survey quarterly. How frequently and how detailed the survey needs to be is dictated by the size and nature of the business and the objectives of the Steering Committee. Similarly, decisions about whether surveys should be conducted by mail or phone, by in-house employees, or by outside consultants should be made by the Steering Committee.

User Groups and Customer Panels

Any organization whose products have user groups or associations should take advantage of the wealth of customer feedback that flows from these groups. Steering Committee members should establish a positive relationship for mutually beneficial interaction with these groups. The Steering Committee might also consider establishing a customer panel which would meet at regular intervals for the purpose of providing feedback to the company.

PUTTING CUSTOMER FEEDBACK TO USE

Customer feedback can be put to use on two levels. First, at the Steering Committee level, feedback can be used as input for the Steering Committee's PDCA cycle. Customer feedback will let the Steering Committee know whether the changes being made are producing the desired results from the customer's perspective. Reviewing customer feedback represents the *Check* phase of the PDCA cycle, and the actions taken as a result represent the *Adjust* phase of the cycle. The Steering Committee should seek to make adjustments that will result in favorable customer feedback. Note that adjustments can be anything from minor or major changes in direction to the formation of new teams to address projects not previously considered.

The second use for customer feedback—and this is by no means less important than the first—is at the project team level. The various project teams can use customer feedback to ratify their activities in a manner similar to that of the Steering Committee. Customer feedback should be studied (the *Check* phase of the teams' PDCA cycles) and used to make adjustments as required (the *Adjust* phase). Admittedly, this can be difficult at the team level. Customer input may not be specific enough for teams to know with certainty that it applies to one process as opposed to another. Judgment must be applied. This is one reason the Steering Committee must stay in close contact with teams. The Steering Committee may be in the best position to judge how customer feedback applies to a particular team's efforts. However, in general, customer feedback can yield information that dictates changes to processes, more robust designs, and the addition of desired features, all of which are traceable to the team level.

The key point here is that customer feedback is a valuable source of information. Although such information is typically available in one form or another, many organizations choose to ignore it. No company can call itself a Total Quality organization unless it has a strong, proactive customer focus. For an organization to realize the full benefits of Total Quality, it must make maximum use of customer feedback.

THE CUSTOMER-FEEDBACK LOOP AS AN ONGOING PROCESS

As long as an organization is in the business of serving customers, the customer-feedback loop must continue. In Step 11 we talked about establishing a customer-satisfaction baseline to identify the starting point from which to measure progress as

Total Quality takes hold. In this step the reader should see customer-satisfaction information as contributing to two processes. The first is the initial follow-up assessment of customer satisfaction after a period of Total Quality activities. The second is ongoing customer input for the PDCA cycles of the Steering Committee and the various teams. This process goes on forever.

EXTERNAL VERSUS INTERNAL CUSTOMER FEEDBACK

This step has been described with the external customer in mind. In the next step we address the issue of internal customer feedback. Most of the points made in this step can also be applied to the internal customer. Keep in mind, though, that only the external customer can vote on the success of an organization's services and products.

SUMMARY

1. It is important to collect and use customer feedback because customers decide the fate of an organization when they decide to use or not use its products or services.
2. Sources of customer satisfaction include warranty data, sales results, Customer Service departments, distributors, unsolicited customer feedback, customer visits, customer-satisfaction surveys, and user groups and customer panels.
3. Customer feedback can be used at two levels: at the Steering Committee level and at the project team level. Both use customer feedback in the *Check* and *Adjust* phases of their respective PDCA cycles.
4. In a Total Quality organization, the customer-feedback loop goes on forever.

KEY TERMS AND CONCEPTS

Customer feedback

Customer panels

Customer-satisfaction information

Customer-satisfaction survey

Customer visits

Distributors

Sales results

Unsolicited customer feedback

User groups

Warranty data

REVIEW QUESTIONS

1. Explain the rationale for collecting and using customer feedback.
2. Describe four common sources of customer feedback.
3. How frequently should the Steering Committee receive customer feedback?
4. How is customer feedback put to use in a Total Quality organization?
5. What is the duration of the customer-feedback loop? Why?

ENDNOTES

1. Robert L. Shook. *Turnaround: The New Ford Motor Company* (New York: Prentice Hall, 1990), 168.

CASE STUDY 18–1

Obtaining Customer Feedback at MTC

By now, MTC's Total Quality implementation had moved well down the road and, in spite of occasional potholes, was working well. Ten teams were working on various projects and another six that had completed their projects. From the point of view of Steering Committee members, positive changes were being made. But were the changes making a difference with customers? This is what John Lee and the other members of MTC's Steering Committee wanted to know.

The top managers at MTC had always monitored warranty data. Recently warranty-related costs had begun an encouraging downward trend. Sales trends were also encouraging. Both warranty and sales information represented useful customer feedback. However, the Steering Committee members wanted more. Consequently, they decided to establish two new teams. The first team would be assigned the tasks of developing a customer-satisfaction survey, periodically conducting the survey, and reporting the results to the Steering Committee. The second team would be given the assignment of forming and periodically meeting with a customer panel. The information provided by these sources, added to that already being monitored, would allow MTC's Steering Committee to carry out its PDCA cycle. The information would also be made available to teams as appropriate.

John Lee was pleased. MTC was coming out of the doldrums and getting on track. Total Quality was working! As he listened to presentations from MTC's various teams each week, John Lee felt as if he might burst with pride. MTC was rapidly becoming a winner, and it was because Lee had taken the bold step of insisting on the implementation of Total Quality.

CASE STUDY 18–2

Customer-Feedback Problems at ESC

To call the next meeting with the Steering Committee tense would be a gross understatement. The discussion of customer feedback among Lane Watkins and his colleagues soon escalated to a battle royal. Not only did the Committee reject his customer-satisfaction survey instrument, they rejected the very idea of a survey. The team leader's opening comment had been, "We aren't going to send out a questionnaire to have our customers tell us what we already know."

The Steering Committee had also rejected Watkins ideas of summarizing customer-complaint information monthly and forming a customer panel. Clearly his colleagues did not intend to pursue the implementation further. "Alright, fine," thought Watkins. "I'll form my own customer panel. There are enough local customers that I should be able to get a group together once a month over lunch."

And that's exactly what Watkins did. He contacted eight local customers and asked each to name a representative to ESC's new customer panel. Six companies agreed to honor his request and an initial meeting date was set. The most enthusiastic customer had been MTC, a manufacturer that until a year ago had used ESC's services frequently. The MTC representative told Watkins that his company was implementing Total Quality too, and had just formed its own customer panel. John Lee complimented Watkins on his idea and said he was looking forward to their first meeting.

Collect and Use Employee Feedback

- Rationale for Requiring Employee Feedback
- The Value of Employee Feedback to Managers, Employees, and the Steering Committee
- Methods for Obtaining Employee Feedback
- Putting Employee Feedback to Use
- Employee Feedback as an Ongoing Process
- Cautions Concerning Employee Feedback

This step is closely related to Step 10, which dealt with identifying employee attitudes and levels of job satisfaction. In Step 10 we established a baseline from which to measure improvement in employee attitudes and job satisfaction as the Total Quality changes were implemented. In this step, we begin a never-ending process of obtaining employee feedback—invaluable information that will help achieve the organizational vision. Since Steps 10 and 19 are so closely related, and since much of what was said in Step 10 about the need for and the process of obtaining employee feedback is applicable here, we suggest that the reader review Step 10 before proceeding.

We have now arrived at the fourth step of the execution phase of the implementation. Some teams have been deployed and more are coming on line. The organization is being managed by the Steering Committee working as a team rather than by individual staff members who are more concerned with their own departments than with the overall organization. The Steering Committee is set to receive feedback from the deployed teams and from customers. Now there is one more constituent group to be heard from: the employees at large.

RATIONALE FOR REQUIRING EMPLOYEE FEEDBACK

In the last step we discussed the importance of collecting and using customer feedback. No Total Quality organization can function without it. Neither can such an organization

TOTAL QUALITY TIP

Talk to Employees and Hear What They Say

Tom Peters, on talking about getting feedback from frontline employees says, "I guarantee you that a lengthy discussion of the 'little things' that rule their lives in your organization will collectively reveal the strategic stumbling blocks to higher quality, more responsiveness, etc."[1]

Tom Peters

function without employee feedback. Customer feedback lets the Steering Committee know how well its organization's products or services are meeting customer expectations. Employee feedback provides another critical type of information: how well the organization's people, tools, and processes are doing in the production of the organization's products or provision of its services. Clearly, an organization must have both kinds of information if it intends to compete in the global marketplace. Customers tell the organization whether its product or service is suitable, acceptable, and worthy of their purchase. Employees tell the organization *why* the product or service is the way it is. The customers' feedback may sound an alarm to tell the organization it needs to find out what is wrong. The employees' feedback may supply the answer.

THE VALUE OF EMPLOYEE FEEDBACK TO MANAGERS, EMPLOYEES, AND THE STEERING COMMITTEE

In our experience, managers who just don't care if their employees are satisfied are in the minority. Most managers are interested in the attitudes and job satisfaction of their employees. They would certainly prefer an organization full of satisfied people with positive attitudes to a workforce full of disgruntled employees. The irony is that although most managers are interested and want satisfied, positive employees, very few really know much about the job satisfaction of their employees. The reason for this is that many managers mistakenly believe that they can know intuitively whether their employees are satisfied or dissatisfied.

The fact is, front-line employees seldom volunteer their real attitudes or job satisfaction level to their managers. Their peers know how they feel, but quite often their managers do not. Consequently, when formal attitude surveys are concluded, managers are sometimes surprised by the results. Yet these same managers typically think they have a good handle on the level of their employees' job satisfaction. The point we want to make here is that, without a concerted effort to measure the level of employee satisfaction, managers are likely to have a very distorted view of how their employees feel. In a Total Quality organization, managers must deal with facts, not supposition. Consequently, managers must know if their employees are satisfied, and, if not, why not. The *why* means that organizations have to go beyond just measuring attitudes; they must also determine the causes behind any dissatisfaction discovered. In doing so, they will find the information needed to improve their organization's processes, products, and services.

Nearly all employees want to do a good job. Nearly all want to be part of an organization they can be proud of. Employees become dissatisfied when the way they are required to do their job is not the best way, or when, no matter how hard they try, the outcome of their efforts is less than it should be. When employees see unacceptable products going to customers, it is difficult for them to take pride in their work or in their organization. The fact that they believe there is a better way and that the product could be improved means they have ideas that could be of value, if only someone would listen.

Without employee feedback, the Steering Committee is flying blind. From employee feedback come the projects that need to be addressed by teams; the ideas that will lead to solutions to problems, improvement to processes and products or services; the data required for effective use of the PDCA cycle; and finally, the ratification of changes that are being implemented. Clearly employee feedback is needed. The question is how to get it.

METHODS FOR OBTAINING EMPLOYEE FEEDBACK

The primary sources of employee feedback are the employees themselves, management, the organization's internal customers, and Human Resources records.

Formal Surveys

The direct approach is often the best approach. Ask for the information. This is what was done in Step 10, when we recommended a formal survey of all employees to establish a baseline of employee attitudes and job satisfaction. Figure 10–1 was provided as a generic survey form that could be tailored to a specific organization. Approximately twelve months after conducting the first survey, conduct another survey using the same survey instrument. Compare the results of the second survey with those from the original survey to see if employee attitudes have changed for the better. The new survey will reveal where progress is being made and where more attention needs to be applied. As in Step 10, the survey can be conducted in-house or by a third party. Regardless of which option was used in conducting the original survey, it is best to use the same option the second time.

The third method explained in Step 10, the Steering Committee discussion, is the least accurate option because it represents the views of Steering Committee members, not real employee feedback. If you used this method to establish baseline data, use it again, but be sure that all members of the Steering Committee strive for objectivity.

Employee attitude surveys should be repeated at regular intervals from now on. We suggest conducting a survey annually for the first several years. After that, an 18-month cycle should be sufficient.

Management Sensing and Listening

With formal survey data being collected at 18-month intervals, in the interim Steering Committee members and all other managers will need to develop and use skills for sensing employee attitudes. The best sensing skill is active listening. There are many different forms of employee feedback available to managers who are good listeners. Meetings are an excellent source of employee feedback for managers. Smaller groups may meet with Steering Committee members to discuss issues, perhaps over lunch or breakfast.

Management by Walking Around is a technique that all managers should practice. While on their rounds, managers should engage in conversation with as many employees as possible. The important thing is to ask questions and listen to the answers. Questions should include such thoughts as, "How can I help make your job easier?" "What can we (management, the Steering Committee) do to help you?" "Are your work instructions and processes as good as they can be?" "What problems do you have that I can help you with?" Whatever the questions are, they should have substance and should be asked in such a way as to solicit feedback on attitudes, job satisfaction, and ideas for improvement.

One might wonder why managers can't simply use a suggestion box. Both techniques are designed to garner employee feedback. However, suggestion systems don't always work as well as they need to, usually because they are used improperly. Well-designed suggestion systems can be an excellent source of employee feedback. However, even the best suggestion system should not be seen as a substitute for walking around, asking, and listening.

Internal-Customer Input

Another element of job satisfaction is how well employees are supported by their internal suppliers. Every employee has *internal suppliers* and is therefore an *internal customer* of that supplier. Along the list of processes an organization employs to manufacture its product or provide its service, there is a chain of internal supplier/customer relationships. How well or how poorly these relationships work not only determines the quality of the organization's output to its ultimate customers, but also impacts the jobs of everyone in the organization. When relationships cause a process to perform poorly, the internal customer of that process has a more difficult task, perhaps having to make corrections, rework, or wait for the next *good* part to come along. The effect ripples throughout the production stream. In the everyday operations of most organizations, the performance of processes can fluctuate. From a statistical perspective, they may be out of control. Sometimes processes can be so unreliable that downstream internal customers are forced to work much harder than they should in order to make the final product acceptable. Hence, the information stored in the minds of internal customers can be invaluable in solving process problems. But it won't come out without a concerted effort on the part of management to solicit their feedback.

The Steering Committee should establish a team to survey internal customers. Ideally, the internal suppliers should conduct the surveys and discuss the issues with their internal customers. This will reveal very useful information not only to these parties, but

---

TOTAL QUALITY TIP

Ultimately, Employees Make an Organization Successful

" . . . some senior managers cannot bring themselves to accept that the success of the company does not rest on how smart they are but on how smart and capable their workers are."[2]

Lloyd Dobyns and Clare Crawford-Mason

---

also to Steering Committee members, who can then deploy teams to pursue problems that surface. Remember that one of the basic tenets of Total Quality is that the people who are closest to the processes are the ones most likely to know about problems. Further, they are the ones most likely to be able to fix them, or to supply vital information to someone who can. Internal customers and their internal suppliers are these people. Managers may not even be aware of everyday production problems—and they never will be without feedback from internal customers.

Human Resources Records

In addition to the type of employee feedback already discussed, valuable information can be gleaned from human resources information such as absenteeism, turnover, and complaints of various kinds. How many employees participate in company-sponsored activities? What do job performance appraisals reveal? Problems in any of these areas usually mean that something needs to be rectified.

PUTTING EMPLOYEE FEEDBACK TO USE

It makes no sense to collect information and then ignore it. Nevertheless, this does happen. Many organizations have more information than they can deal with. Some find it difficult to separate the important from the trivial, with the result that trivial matters divert attention and resources away from important ones. We raise this issue because your organization may be faced with such a situation. There is probably not a more difficult area in which to separate the significant from the trivial than the human resource component of the organization. This poses a problem when soliciting employee feedback.

The Steering Committee has to make a concerted effort to acquire feedback, sort it out in terms of what is most important and what can be left alone for the time being, and then take action on all input. This is not a contradiction. All suggestions should be responded to, even those that cannot be accepted. This involves letting the employees who provided the suggestion know that it was considered and explaining why the suggestion was not put in place. Feedback to employees is critical. Without it, your sources of employee input will dry up quickly.

The Steering Committee and all project teams that are deployed need to make use of the appropriate statistical tools to determine what is most important (Pareto charts are a valuable tool here), and to find the root causes of the problems they are trying to solve. Over time, all of the tools will be useful. Even in the earliest stages you will find Pareto charts, fishbone diagrams, stratification, flow charts, and surveys to be helpful tools. The other tools—check sheets, histograms, scatter diagrams, run charts, and control charts—should be applied as appropriate for dealing with the problem in question. In addition, the Steering Committee and teams should keep the PDCA cycle in mind.

Employee feedback is simply another source of information that is meant to help the Steering Committee identify projects and then to furnish the *Check* information for the project's Plan/Do/Check/Adjust cycle. If the *planned* actions worked as anticipated when it was *done*, then the *check* information from employee feedback should have verified it. Or perhaps the *check* information indicated that the action taken had not worked as planned, leading the Steering Committee to *adjust* the action for the next cycle.

To summarize, employee feedback is available in many forms, including the following:

Survey data	Performance data
Attitudes	Comments
Absenteeism rates	Job satisfaction
Suggestions	Complaints
Turnover	

Employee feedback is valuable input for the following:

Identifying issues	Selecting projects
Identifying solutions	Measuring progress
Setting tactical goals	Carrying out the PDCA cycle

EMPLOYEE FEEDBACK AS AN ONGOING PROCESS

Like all of the tasks in the planning and execution phases of Total Quality, obtaining employee feedback and acting on it are never-ending functions. Once again, it is the responsibility of the Steering Committee to ensure that this function is accomplished. Informal feedback should be solicited continually, from all sources available. We cannot overstate the importance of using feedback from employees. Failing to use this vital information can cause the implementation to fail.

CAUTIONS CONCERNING EMPLOYEE FEEDBACK

Steering Committee members and all other managers must guard against the "shoot the messenger" syndrome. Most people don't like to receive bad news, but wise managers understand that the sooner they know about the bad news, the better. Managers whose response to bad news is to shoot the messenger will soon find that they receive no news at all. This happens at all levels in an organization. For example, the vice presidents of one major U.S. corporation worked well with their CEO and felt free to convey any message to him. When this general manager moved to a new assignment, he was replaced with a man who ridiculed any ideas that did not coincide with his. He also tended to shoot the messenger bearing bad news. Predictably, the staff soon became reluctant to tell him anything. As a result, the business went into decline and eventually failed. If managers want to hear people's ideas and concerns, they must accept them in a nonjudgmental manner. Otherwise, they will be cut off from what could be their most valuable source of information and ideas.

Confidentiality is another issue of great concern when collecting employee feedback. Surveys administered to employees should be designed to protect the respondent's anonymity. Further, managers should make it clear to employees that they have no interest in knowing who said what, that honest responses are crucial to results, and that only by giving honest responses can they help the organization improve. Make sure that employees know there will be no retribution for telling the truth. Failure to make these guarantees may result in no feedback, or feedback of no value.

SUMMARY

1. The rationale for soliciting employee feedback is that it provides vital information about how well the organization's people, tools, and processes are working.
2. Managers should monitor employee attitudes constantly by walking around, questioning, and listening, and by conducting formal surveys.
3. Feedback that is collected should be used. In addition, all employee input should receive a prompt response.
4. Employee feedback is available in many different forms, including survey data, performance data, attitudes, comments, absenteeism and turnover rates, suggestions, job satisfaction, and complaints.
5. Employee feedback can be used for identifying issues, selecting projects, identifying solutions, measuring progress, setting tactical goals, and carrying out the PDCA cycle.
6. The process of soliciting and using employee feedback is ongoing.
7. Managers should make a special effort to avoid the "shoot the messenger" syndrome. Managers who punish good employees for delivering bad news soon receive no news at all. Employees who offer feedback should be given a guarantee of confidentiality.

KEY TERMS AND CONCEPTS

Attitudes	Employee attitudes
Comments	Employee feedback
Complaints	Formal surveys
Confidentiality	Identifying issues

Internal-customer input   Sensing employee attitudes

Job satisfaction   Setting tactical goals

Measuring progress   "Shoot the messenger" syndrome

PDCA input   Suggestions

Selecting projects   Turnover

REVIEW QUESTIONS

1.   Explain the rationale for requiring employee feedback.
2.   Describe the various ways employee feedback can be obtained.
3.   Describe the various ways employee feedback can be put to use.
4.   What are two cautions relating to the collection of employee feedback? Explain them.

ENDNOTES

1.   Tom Peters, *Thriving on Chaos* (New York: Harper-Collins Publishers, 1991), 529.
2.   Lloyd Dobyns & Clare Crawford-Mason, *Thinking About Quality, Progress, Wisdom, and the Deming Philosophy* (New York: Times Books, A Division of Random House, 1994), 23.

CASE STUDY 19–1

Collecting Employee Feedback at MTC

Now that the Total Quality approach to doing business had caught on at MTC, John Lee was anxious to make sure there was no backsliding. Employee feedback was what he and the other members of the Steering Committee wanted, and they were getting it. A formal survey had already been conducted. The data had been summarized, analyzed, and was now being acted on. Feedback from the survey had resulted in the establishment of three new teams.

Now John Lee and the other Steering Committee members conducted at least one walking tour of the organization every working day, talking and listening to employees. These walking tours served at least three purposes. First, they got the committee members out of the seclusion of their offices and into the places where the work was done. Second, their presence showed both interest and commitment. Third, the feedback they received was given firsthand and on a one-on-one basis. Finally, the walking tours helped Steering Committee members get to know their employees better, and vice versa. John Lee thought he could feed the comfort level of employees improving every day. And the more comfortable they got, the better their feedback became.

The next step in John Lee's mind was the implementation of chalk talks with internal-customer suppliers. The chalk talks would involve internal customers sitting down in a conference room with their suppliers and diagramming on a chalkboard ways to improve their interaction. Lee's hope was that eventually these chalk talks would become spontaneous, requiring no prompting by management.

===== **CASE STUDY 19–2** =====

The Sad Story of Implementation Gone Awry at ESC

The first meeting of ESC's customer panel had gone well and badly. Lane Watkins thought it had gone well in that the panel members had provided him with excellent feedback. Of course, Watkins doubted he could convince higher management at ESC to use it, but that was a worry for another day. The meeting had gone badly in that John Lee, the panel member representing MTC, had asked some probing questions. Before the meeting had ended, Watkins had become convinced that Lee saw right through ESC's thin veneer of Total Quality.

Consequently, Watkins hadn't been surprised when John Lee stayed behind after the meeting broke up. After the other panel members had departed, Lee had approached Watkins and asked point-blank, "Do you want to tell me what's really going on?" Watkins hesitated for just a moment before deciding that getting his problems with this assignment off his chest with someone who understood Total Quality was just what he needed. "Can I buy you supper?" was his response.

Supper turned into a four-hour discussion, but when Watkins left the restaurant he was walking on air. He hadn't felt so good in months. "What a manager!" he thought. "I'd give my right arm to work for that CEO." It had been an interesting night for John Lee, too. "ESC wouldn't be in such bad shape if they'd give that young man and others like him a little support," thought Lee. "What a waste of talent!"

Change the Infrastructure

MAJOR TOPICS

- Rationale for Changing the Organization's Infrastructure
- Identifying and Eliminating Organizational Roadblocks
- The Hierarchical Organization versus the Total Quality Organization
- Examining the Organizational Infrastructure
- Changes to Infrastructure as a Steering Committee Function
- Evaluating Proposed Changes
- Repeating the Total Quality Cycle
- The Total Quality Legacy

With Step 20, we come to the final step in the implementation process. By now, the Steering Committee is managing the organization as a team rather than as a group of individual department heads. The Steering Committee is focused on the organization's vision, guiding principles, and broad objectives rather than on departmental agendas. Teams are deployed and are working on issues selected by the Steering Committee in support of the vision. The teams are providing feedback to the Steering Committee. Changes are taking place: changes in the procedures, changes in processes, and changes in the way employees think of customers—both internal and external. Total Quality is beginning to take root.

However, don't think of Step 20 as the final step in the implementation process. Rather, think of it as the final step *in one cycle* of the process. The cycle continues forever. When you have completed Step 20 the first time, you have done all of the implementation steps once. Consequently, in succeeding cycles, all of the steps will become more and more natural until, finally, they become the way your organization operates.

RATIONALE FOR CHANGING THE ORGANIZATION'S INFRASTRUCTURE

Before discussing why changes to the infrastructure should be made, we need to define the term as it applies to an organization. Webster's dictionary defines *infrastructure* as:

> "a sub-structure or underlying foundation; esp., the basic installations and facilities on which the *continuance and growth* of a community, state, etc. *depend*. . . ." [italics added][1]

In our case, we are talking about the infrastructure of a business or any other kind of organization. An organization's infrastructure certainly includes the buildings which house the workspace, the heating and cooling systems, and so on. But an organization's infrastructure also includes its organizational structure; business systems, processes, and procedures; union rules; award and recognition programs; and even executive perks. These are the parts of the infrastructure with which we are most concerned as we implement Total Quality.

The organization's hierarchy, if it is typical of most traditional structures, will be functionally oriented and will have too many layers, which can impede efficiency and contribute to waste. As root causes to operating problems are found, many will point to the organizational chart, and specifically to the barriers between departments and between the people who are trying to do the work. A similar situation often exists with the organization's operating procedures. They tend to be tied to the old, traditional way of doing things, and therefore stand in the way of improvements in quality and efficiency. As teams examine the organization's procedures and processes, they will inevitably find that changes need to be made to improve quality, consistency, and efficiency.

The rationale for changing the organizational infrastructure, then, is that Total Quality requires it. If the organization is to become a competitive, efficient, customer-oriented agency able to succeed in a global environment, change will be required and will become a way of life as new opportunities are discovered by the teams. Basically, an organization can accept change, or it can stay as it is until it no longer has a reason to exist. Going back to the definition of infrastructure, it is the underlying foundation on which the *continuance and growth* of the organization depend. Given today's competitive and cost-conscious environment, too many organizations have a foundation that cannot support continuance and growth.

IDENTIFYING AND ELIMINATING ORGANIZATIONAL ROADBLOCKS

The following examples are provided to illustrate the types of organizational roadblocks that are likely to be encountered and ways to eliminate them. The term *roadblock* as used here means any impediment to improvement that cannot be resolved by the persons directly involved.

For our first example, we return to the company that wanted to improve the efficiency of the transition from design to production for its new products. A team was formed by the Steering Committee and asked to find ways to eliminate the production problems that usually occurred after the Engineering Department had completed its work and forwarded its designs to manufacturing. Production problems were costly because they often required redesign and manufacturing rework. This, in turn, increased the time-to-market period, which gave the competition a head start.

The team was composed of employees who were directly involved with the work in question. In a short time, the team concluded that the best way to solve the problem was to involve manufacturing personnel in the design process from the beginning of the project. They found that many companies were doing this successfully and calling it *concurrent engineering* or *design for manufacture (DFM)*. By having manufacturing personnel involved from the start, anything that would cause production problems would be identified during the design process rather than afterwards. Manufacturing personnel could suggest changes to eliminate the problem, reduce manufacturing costs, and reduce cycle time.

When the team made its presentation to the Steering Committee, its recommendation was seen as a logical solution to the problem. However, the company was not structured for cross-functional teams. Engineering and Manufacturing were two distinct departments, each with its own senior manager. Engineering was accustomed to creating a design and handing it over the wall to Manufacturing. Manufacturing would then make the product. This was the way it had always been, and if the company had not become involved in Total Quality, this is the way it would have stayed. The employees directly involved with the problem could do nothing about it. But as members of the Steering Committee, working together to accomplish the vision, the heads of the Engineering and Manufacturing departments, along with the rest of the team, agreed to try the proposed solution.

At first it was only a glimmer of an idea, but gradually the same question formed in several of the Steering Committee members' minds: *Why are engineering and manufacturing two separate departments?* This led to a discussion in which Steering Committee members concluded that a lot of the current barriers to communication, cooperation, and overall efficiency would be eliminated if the two departments were combined. One Steering Committee member suggested that it might be smart to organize the entire company around its four major product lines. It would be easier to focus on customers if each product line had its own engineering, manufacturing, sales, and service personnel. Instead of having departments divided by function, there would be a single, composite group for each product line. Some companies had adopted this approach with excellent results.

In this first example, the Steering Committee made a commitment to change the way products are designed, and then changed the organization's structure to combine two departments and thereby eliminate barriers. Next, the Steering Committee began considering a change in organizational philosophy to promote the concept of customer focus. These changes are all infrastructure changes.

The second example involves a Steering Committee that established a cross-functional team to reduce rework in manufactured subassemblies. The team developed a flow diagram of the process and found that the subassemblies—small electronic circuit boards—were batch-processed fifty at a time. Several batches would accumulate before an inspector examined them for defects. The inspector would then send the defective circuit boards to a rework area where several highly skilled operators repaired them. At this point the circuit boards were passed to the next step in the process. The team identified the following problems with the process:

1. Inspections were too infrequent, allowing errors to be made repeatedly before being caught.
2. Feedback to the person making the error was too far removed in time from when the error was committed, or was not reported to the individual at all.
3. Because a third party made the repair, the person who made the error was out of the loop, and consequently was likely to repeat the error continually.

The team made the following recommendations to the Steering Committee for solving the problems:

1. Have inspections made by the operator manning the succeeding process rather than by the inspector, and include the inspection as a part of the process rather than in large batches.
2. Report any errors to the person responsible, and have that person do the rework.
3. Provide training (i.e., in inspection and rework) for the people involved.
4. Have the inspector perform audits of products and procedures to confirm compliance with procedures, rather than routinely check for errors.

The team identified several Total Quality companies that were successfully using techniques similar to those being recommended. Inspectors in these companies were doing higher level work, such as audits, and the employees who once made repeated errors were now able to eliminate them.

The Steering Committee in this example adopted the team's recommendations, thus setting the wheels in motion to update organizational procedures to reflect the changes.

In a traditional organization there is always the voice proclaiming, "This is the way we do it because this is the way we've always done it. Our procedures are what they are for a reason." Actually, this is true. However, the reason the procedures haven't been changed is that anyone who thought of a better way was squelched with the "this is the way we've always done it" argument. This doesn't happen in a Total Quality organization. Procedures are considered targets for improvement just like anything else.

THE HIERARCHICAL ORGANIZATION VERSUS THE TOTAL QUALITY ORGANIZATION

Figure 20–1 contrasts the traditional hierarchical organization to the structure of an organization that has implemented Total Quality.

Most Western organizations, both public and private, are designed in the traditional hierarchical mode. That is, the organizational chart resembles a pyramid, with the person in charge at the peak. At the bottom are the people who do the work the organiza-

Figure 20–1
The Hierarchical Organization
versus the Total Quality
Organization

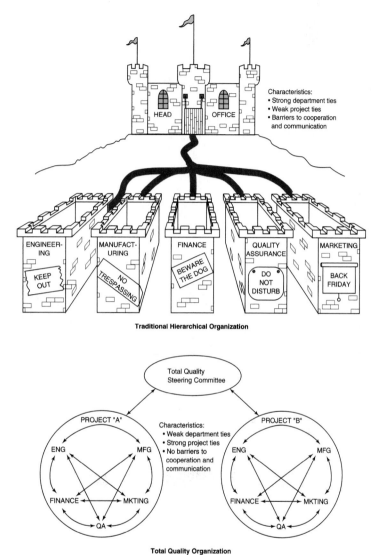

Characteristics:
• Strong department ties
• Weak project ties
• Barriers to cooperation and communication

HEAD OFFICE

ENGINEER-ING

MANUFACT-URING

FINANCE

QUALITY ASSURANCE

MARKETING

KEEP OUT

NO TRESPASSING

BEWARE THE DOG

DO NOT DISTURB

BACK FRIDAY

Traditional Hierarchical Organization

Total Quality Steering Committee

PROJECT "A"

Characteristics:
• Weak department ties
• Strong project ties
• No barriers to cooperation and communication

PROJECT "B"

ENG   MFG

FINANCE   MKTING

QA

ENG   MFG

FINANCE   MKTING

QA

Total Quality Organization

tion is chartered to do. How many layers there are in a pyramid varies widely from organization to organization. Over time, as organizations grow, the pyramid tends to become taller and taller. There are problems with this type of structure.

■ The taller the structure (the more levels it has), the more difficult it becomes to communicate, both up and down the hierarchy. As messages move through the organization through intervening layers, the individuals involved interpret the information, putting their own slants or biases into the communication process and dis-

---

TOTAL QUALITY TIP

Do Hierarchies Dampen the Human Spirit?

"The hierarchial structure where everyone has a superior and everyone has an inferior surely is corrupting of the human spirit—no matter how well it served us during the industrial period."[2]

John Naisbitt and Patricia Aburdene

---

torting the original message. Moreover, with so many layers to go through, communication becomes extremely slow.

- With the organization split up by departments, each with its own management, communication—and consequently cooperation—among people in different departments becomes difficult.

- Individual departments, each with its own agenda and operating rules, tend to become an end unto themselves. As a result, departments tend to complete with one another and pull in different directions. Interdepartmental relationships become adversarial, to the detriment of the total organization.

- Over time, as departments attempt to become self-sufficient, or at least nondependent on one another, inefficiency creeps into the organization.

- Allegiance is often to the department rather than to the overall organization. Customer focus is lost; only the sales department may recognize the customer at all.

- The hierarchical organization is based on the *superior/subordinate* relationship. That some are *superior* implies that others are *inferior*. *Subordinate* is just a nicer way of saying *inferior*, and everyone knows it.

This list could go on, but the point is, the type of organization we have described is out of step with Total Quality. Eventually, one of the teams will bring up the problem, or a piece of it. They will conclude that their project can be successful only if some structural change is made. When this recommendation is brought to the Steering Committee, the Committee will have to decide whether the organization should remain as it is or be transformed, perhaps a piece at a time, to conform to the Total Quality philosophy. As the changes are made, others will come into focus. A typical result is the elimination of whole layers of management from the structure.

If the Total Quality philosophy is followed, eventually the organization will change from tall to broad. Where managers are employed, they will have a much broader span of control, thereby increasing the number of people who report to them. In traditional organizations, managers could not cope with a large number of employees vying for their attention. Under Total Quality, employees are empowered to solve most of their own problems. Consequently, managers can supervise more employees.

EXAMINING THE ORGANIZATIONAL INFRASTRUCTURE

Two questions that should be asked about every part of the organization's infrastructure are: How did it get this way? Does it still make sense? Allowing any preexisting function to block a possible improvement must be considered a cardinal sin. Yet this happens with regularity in traditional organizations. Worthwhile changes are prevented in the name of familiarity, tradition, and fear. In a Total Quality organization, every aspect of the infrastructure should be challengeable. There are no sacred cows. No part of the organization is considered sacrosanct.

One of the most difficult aspects of the infrastructure for many companies to deal with is the question of executive perquisites, or perks. Yet even these should be challenged because they can send the wrong message to the majority of employees who do not get the perks. For example, some larger companies have executive dining rooms for their senior managers. The meals typically are free and are better than those served in the cafeteria, where employees who earn far less must pay for their meals. When executive perks such as dining rooms were first established, employees may have thought little about the issue. However, times are different now.

While they should be compensated appropriately, managers should not enjoy major perks that substantially separate them from employees. Employees are important. They have contributions to make and they should take on a larger role in the operation of the organization. Managers are there to serve, to help, to facilitate. This is what Total Quality is all about, but it cannot happen if managers segregate themselves from employees. In today's environment, ostentatious executive perks are completely out of place. The message such perks convey is that, while managers may talk about participation, empowerment, and facilitating, they still consider themselves to be above the rest of the employees. As Total Quality organizations try to bring managers and employees who have traditionally been separated closer and closer together, many types of executive perks are coming under close scrutiny.

Many other features of an organization will become targets of change. When practices have been in place for a long time, proposals to change them are likely to be met with resistance. Employees may be so comfortable with certain ways of doing things that they do not want to see them changed. But comfort is no reason to perpetuate a practice. The only valid reason for ever maintaining the status quo is that the status quo is still the most competitive approach. If a proposed change makes sense, it should be implemented, period. As the Total Quality organization works toward continuous improvement, existing procedures and processes are certain to become candidates for change. It is also certain that some people will attempt to prevent needed changes by invoking tradition. Do not accept tradition or any other roadblock as an excuse for not making a change that will benefit the organization.

CHANGES TO INFRASTRUCTURE AS A STEERING COMMITTEE FUNCTION

Often proposed changes to the infrastructure can be made only by the Steering Committee. Changes to the organizational structure such as the addition or elimination of capital equipment or facilities, work rules, and award/recognition programs require Steering Committee approval. Of course, the Steering Committee might authorize a project team to make changes to the line-level processes. In any case, the Steering Committee as the senior management entity is still responsible for the change, and must be fully informed and actively involved in the decision to proceed or not.

As members of the Steering Committee, senior managers retain full responsibility for the well-being of the organization and for the best interests of the organization's stakeholders. Total Quality managers understand that personal agendas or departmental agendas must be subordinated to the overall vision of the organization. They see their leadership as more than just a function for which they are compensated. They recognize that their job is to make certain the organization continues as a viable entity and achieves sustained growth. Total Quality managers understand that success in today's competitive, cost-conscious environment is achieved by continuously improving all aspects of the organization's business, including its products, services, business practices, processes, and procedures. Much of what will need to be changed is included in what we have defined as the infrastructure. Total Quality managers have a mandate to challenge every aspect of the infrastructure and to make changes when change will promote achievement of the vision.

EVALUATING PROPOSED CHANGES

Changing the infrastructure can be one of the most difficult tasks encountered in the Total Quality implementation. It is therefore an area that the Steering Committee must keep in the forefront of its thinking, to ensure that when changes are appropriate, they are adopted. For many organizations, the alternative to change is failure. However, change for the sake of change can be just as damaging. Just as you should continually challenge every aspect of the infrastructure, you should also clearly test the viability of proposed changes before adopting them. The cure, in some cases, might be worse than the disease. For example, if a recommendation to downsize the organization is presented, the Steering Committee must test the recommendation to see whether it is necessary, whether the desired results (usually cost cutting) will, in fact, be achieved, or whether the recommendation will only serve to substitute one problem for another. Downsizing has been the trend in major North American and European corporations since the mid-1980s and early 1990s. In some cases, downsizing was the right thing to do. But in other cases, downsizing was an inappropriate response made by managers who lacked the competence to turn their organizations around.

Too often, layoffs are used as a quick way to cut expenses by organizations whose vision extends no further than the next quarterly dividend. When massive layoffs take place, it often means that senior management has failed to live up to its mandate for continued growth. Remember that the mandate for management is for continued prosperity and growth in concert with the organization's long-range vision, guiding principles, and broad objectives. Test the current infrastructure and any proposed changes against these factors, and your organization will be on firm ground.

REPEATING THE TOTAL QUALITY CYCLE

You have now gone through all of the steps of the implementation at least once, but the implementation is not complete with just one pass-through. As we have discussed, Steps 12 through 20 represent a constantly repeating cycle that is the engine for continuous improvement. Step 7, "Communicate and Publicize," also goes on forever. In addition, the other steps should be revisited occasionally to refresh memories and correct any deviations from the planned course. When the Steering Committee and all other employees consider Total Quality to be "the way we do things," when the Total Quality approach has become natural, and when the indicators show that the organization is moving toward accomplishing its vision, then the implementation can be considered complete.

SUMMARY

1. An organization's infrastructure consists of its facilities; organizational structure; business systems, processes, and procedures; union rules; award/recognition programs; and executive perks. In order to change an organization, it is necessary to change its infrastructure. The rationale for changing the organization's infrastructure, where and when necessary, is its continued survival and prosperity.
2. An organizational roadblock is an impediment to improvement that cannot be removed by the persons directly involved. One of the main responsibilities of the Steering Committee is removing organizational roadblocks so that continuous improvement can occur.
3. One of the characteristics of a Total Quality organization is that it has fewer hierarchical levels than traditional organizations. This means that managers in Total Quality organizations typically have more people reporting to them than their counterparts in traditional organizations. Because employees in a Total Quality organization are empowered to make their own decisions and solve their own problems, managers are able to supervise more employees.
4. In a Total Quality organization, there are no sacred cows. Existing practices, processes, and procedures are challenged constantly. Proposed changes are tested before they are implemented. Traditions are not held to for the sake of tradition, nor are changes made simply for the sake of change. Practices and procedures are either retained or changed based on what is most likely to make the organization more competitive.
5. The fact that an organization has gone through all 20 steps of the implementation process does not mean that the implementation is complete. The implementation is not complete until the Total Quality way is the accepted and natural way things are done in the organization, and the indicators show that the organization is making acceptable progress toward achieving its vision.

KEY TERMS AND CONCEPTS

Business systems, processes, and procedures   Executive perks

Continuance and growth   Hierarchical structure

Infrastructure   Sacred cows

Organizational levels   Tradition

Repeating the cycle

REVIEW QUESTIONS

1. Explain the rationale for changing the organization's infrastructure.
2. What are organizational roadblocks? What is the Steering Committee's responsibility with regard to organizational roadblocks?
3. Describe the difference between the structure of the traditional organization and that of the Total Quality organization.
4. When can the Total Quality implementation be considered complete?

ENDNOTES

1. *Webster's New World Dictionary of American English* (New York: Simon and Schuster, Inc., 1988), 694.
2. John Naisbitt and Patricia Aburdene, *Reinventing the Corporation* (New York: Warner Books, Inc., 1985), 41.

CASE STUDY 20–1

Completing the Implementation at MTC

John Lee straightened his tie and walked to the podium. This was a moment he had been anticipating for over two years. As he looked out over the banquet hall, he swelled with pride at what he saw. Every single MTC employee had accepted his invitation to the first annual MTC Family Night Banquet, and many had brought their spouses and children. He smiled and acknowledged their applause. What made this moment especially poignant for Lee was that he knew their applause was sincere. MTC was doing better than it ever had, and the employees appreciated the leadership John Lee had provided in getting things turned around.

After introducing the members of the Steering Committee, all of whom were seated in the audience with the employees and their families, Lee began his speech. It wasn't long, nor particularly eloquent. But to MTC's employees, it was the best speech they had ever heard. This is what Lee said:

> "Two years ago, our company was struggling just to stay in business, and our prospects were bleak. We were down, and there weren't many people in the business who thought we could make it. In fact, the buzzards were already circling. I can still remember crossing the street to avoid a particularly aggressive creditor. But I wasn't ready to throw in the towel, and neither were you. So I asked you to join me and the other members of what later became our Steering Committee in betting our future on Total Quality. You agreed not just to give it a sincere try, but to give it all you had.

"Thanks to your determination and the leadership provided by the Steering Committee, MTC is now a Total Quality Company. Our products are top-notch, our customers are happy, our profits are up, and our costs are down. We have become a team. We are moving toward our vision, operating according to our guiding principles, and accomplishing our objectives. And we continue to improve everything, every day. In other words, your jobs are secure. Thank you for saving our company."

=============== **CASE STUDY 20–2** ===============

A Beginning and an End at ESC

Lane Watkins had to pinch himself to make sure he wasn't dreaming. He had been in his new job at MTC for less than a week now, and he was already a member of a project team. The team's assignment was to find ways to shorten the product-development phase for MTC's new products. After almost two years of frustration at ESC, Watkins felt like a kid in a candy store every time he walked through MTC's front door. He had fit in immediately. It quickly became apparent that every employee at MTC was on a mission to continually improve everything, all the time. Watkins loved the calm confidence and sense of purpose that was ever-present among MTC's employees. While coming to work at ESC for the past two years had been a dreaded experience, coming to work at MTC was a joy.

He was participating in a brainstorming session with his project team when he noticed John Lee quietly slip through the door and take a seat at the back of the room. When the team took a break, Lee joined him at the conference table and asked how things were going. Watkins told him that he had never felt better about his work or his career. Lee congratulated him and thanked him for the excellent work he was already doing for MTC. Then he handed Watkins a copy of the afternoon newspaper, folded to the business page. The headline read: "ESC Files for Bankruptcy."

The Total Quality Legacy

As a final illustration of the advantages of the Total Quality philosophy, we turn to a page in history.

On June 6, 1944, the Allies hit the beaches at Normandy. The dissimilar command protocols of the American and German forces during this historic battle are telling.

Even before the Allied troops landed at Normandy, the meticulously crafted plans for the invasion had unravelled. The invasion fleet expected calm seas, but was instead hit by the worst storm ever recorded for the English Channel during the month of June. Paratroopers were supposed to be dropped at a location from which they could assist in the landing, but clouds obscuring the drop zone caused them to miss the target area by several kilometers. The German pillboxes and bunkers overlooking Omaha Beach were to have been taken out by bombing just prior to the assault, but actual damage to them was negligible. In short, the Allied plan fell apart from the beginning. However, one thing did go according to plan: the Germans were caught off guard. The Germans believed the invasion would come further east, at Calais. Consequently, much of their armor was miles away from the actual invasion site.

American soldiers and their officers were briefed on the mission and taught to think and act on their own. When faced with a situation that was vastly different from what their commanders had expected, they improvised and carried the day.

The behavior of the German forces shows marked contrast. The commander of the closest German Panzer Division could have swiftly moved his mechanized forces to the scene of battle. Had he done so, the outcome could have been much different. But he did not, because German officers were trained to follow orders without questioning and, in the absence of orders, to do nothing. To a German officer, to deviate from the last orders received was unthinkable. The Panzer commander's superior, Field Marshall Erwin Rommel, was in Germany and could not be reached. Adolph Hitler was sleeping and no one dared wake him. The result? The German tanks did not move, and the Americans managed to secure the beachhead.

The American officers and soldiers had been well briefed on their objectives and given the authority to do whatever was necessary to achieve them, while the German officers had not. In other words, the Americans were empowered and their German counterparts were not. Empowerment prevailed in what has been called "the decisive battle of the century."

Often, on the battlefield of commerce, two organizations compete for the same market or objective. When one organization is manned by well-informed employees who are empowered to solve their own problems and to communicate issues freely to managers, while the other is staffed by people who are expected to simply do as they are told, which one will prevail? The record shows that in business, the former organization—the one that follows Total Quality principles—generally prevails. As long ago as 1944, it was shown that these concepts apply not just in business, but on a real battlefield.

INDEX